Joyce Ravid

Sonia Shah is an investigative journalist and the critically acclaimed author of *The Body Hunters: Testing New Drugs on the World's Poorest Patients* and *Crude: The Story of Oil.* Her writing has appeared in *The Washington Post, The Boston Globe, New Scientist, The Nation,* and elsewhere.

ALSO BY SONIA SHAH

Crude: The Story of Oil

The Body Hunters: Testing New Drugs on the World's Poorest Patients

THE FEVER

THE FEVER

HOW MALARIA HAS RULED HUMANKIND FOR 500,000 YEARS

SONIA SHAH

PICADOR

A SARAH CRICHTON BOOK
FARRAR, STRAUS AND GIROUX
NEW YORK

Printed in the United States of America. For information,
address Picador, 175 Fifth Avenue, New York, N.Y. 10010.

www.picadorusa.com

For information on Picador Reading Group Guides, please contact Picador.
E-mail: readinggroupguides@picadorusa.com

Title-page illustration: *A Man Being Attacked by Insects,* from *Hortus Sanitatis,* 1491,
courtesy of the National Library of Medicine

The diagram of malaria on page 243 is courtesy of the
CDC/Alexander J. da Silva, Ph.D./Melanie Moser

Designed by Abby Kagan

The Library of Congress has cataloged the Farrar, Straus and Giroux edition
as follows:

Shah, Sonia.
The fever : how malaria has ruled humankind for 500,000 years / Sonia Shah.—
1st ed.
p. cm.
Includes bibliographical references and index.
ISBN 978-0-374-23001-2
1. Malaria—History. I. Title.
[DNLM: 1. Malaria—history. WC 750 S525f 2010]
RA644.M2S46 2010
614.5'32—dc22

2010002374

Picador ISBN 978-0-312-57301-0

First published in the United States by Sarah Crichton Books,
an imprint of Farrar, Straus and Giroux

D 10 9 8

Man ploughs the sea like a leviathan, he soars through the air like an eagle; his voice circles the world in a moment, his eyes pierce the heavens; he moves mountains, he makes the desert to bloom; he has planted his flag at the north pole and the south; yet millions of men each year are destroyed because they fail to outwit a mosquito.

—PAUL F. RUSSELL, 1931

CONTENTS

1 ▪ MALARIA AT OUR DOORSTEP 3

2 ▪ BIRTH OF A KILLER 11

3 ▪ SWEPT IN MALARIA'S CURRENT 34

4 ▪ MALARIAL ECOLOGIES 59

5 ▪ PHARMACOLOGICAL FAILURE 86

6 ▪ THE KARMA OF MALARIA 121

7 ▪ SCIENTIFIC SOLUTIONS 141

8 ▪ THE DISAPPEARED: HOW MALARIA VANISHED FROM THE WEST 170

9 ▪ THE SPRAY-GUN WAR 193

10 ▪ THE SECRET IN THE MOSQUITO 219

MALARIA DIAGRAM 243

NOTES 245

ACKNOWLEDGMENTS 295

INDEX 297

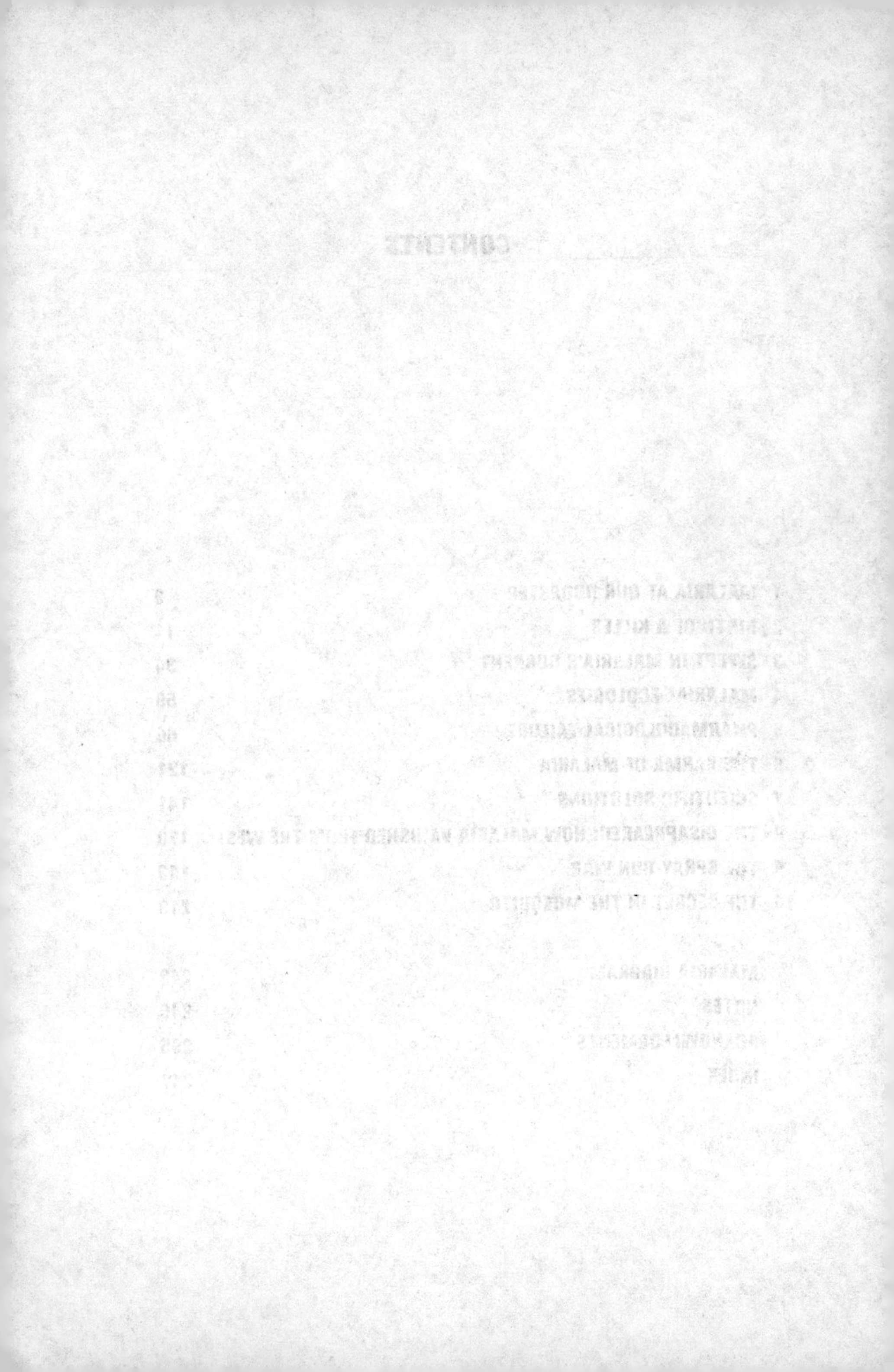

THE FEVER

1. MALARIA AT OUR DOORSTEP

The view through the mosquito net is blurry, but I can see the thick skin of grime on the leading edge of each blade of the ceiling fan as it slowly whirs around, keening alarmingly. This is how it was every summer when I visited my grandmother's house in southern India. While my cousins snore on the bed mats laid across the floor beside me, glistening bodies bathed in the warm night breeze, my sleeping mat is ensconced in a hot, gauzy cage. The mosquitoes descend from the darkened corners of the whitewashed room and perch menacingly on the taut netting, ready to exploit any flicker of movement from their prey within. It is hard to fall asleep knowing they are there, watching me, but eventually I drop off and my tensed body uncurls. They sneak into the gaps my protruding limbs create, and feast.

In the morning, all my hard work of trying to fit in, to overcome the Americanness of my suburban New England life, has been undone, for my Indian cousins are smooth and brown while I am speckled with bleeding scabs. My grandmother vigorously pats talcum powder over my wounds, the white powder caking pink with congealed blood, as my cousins snicker. I don't understand how they escape unscathed while I am tormented. But incomprehension is part of the package of these childhood summers in India. Just outside my grandmother's house ragged families huddle in rubble along the road and use the train tracks as their toilet. They wave their sticklike arms in my face and moan woefully when we pass by on the way to temple, caricatures of beggars. One boy's leg has swollen to the size of a log, and is gray and pimpled, from some disease brought on by a mosquito bite. My grandmother tightens her grip on my hand. We give the children nothing. I can't understand this, either. When we get to the white marble temple, it is full of incense and golden statues encrusted with diamonds and rubies—to my seven-year-old mind, the very picture of prosperity.

Part of me despises my estrangement, my incomprehension, the fact that I must sleep under the suffocating net and take the malaria pills while my cousins don't. But part of me is secretly glad. The boy with the swollen leg frightens me. The family who lives on the curb frightens me. India frightens me. These fears, for the girl who is supposed to be Indian but isn't, are unspeakable.

When no one is looking, I crush the mosquitoes' poised little figures with my palm and smear the remains on a hidden seam in the couch. Our Jain religion forbids violence of any kind. No eating meat. No swatting flies. My grandmother wears a mask over her mouth while she prays, to protect airborne microbes from inadvertent annihilation in her inhalations, and considers walking on blades of grass a sin. Meanwhile, there I am in the corner, cravenly pulverizing mosquito corpses behind my back, blood literally on my hands.

Back home in New England, the mosquitoes still bite, but there are no nets at night, no pills to take, no scary beggars on the side of the road. We shop for forgettable plastic trinkets at the mall. My fear and loathing of the mosquito are blunted into games of tag. My father calls himself Giant Mosquito, undulates his fingers like proboscises and chases me and my sister. It's scary, but fun-scary. We screech with glee and stampede through the house.

• • •

Thirty years later, on the S-shaped land bridge between the North and South American continents, I meet José Calzada. Calzada is a mosquito stalker of sorts, and I, the mosquito hater, have come to learn about the local mosquitoes and their exploits. A parasitologist from Panama City, Panama, Calzada spends his time rushing to the scene of disease outbreaks across the isthmus. The mosquito-borne parasite that causes malaria, *Plasmodium*, is one of his specialties.

It is April 2006. For most of the past century, there hasn't been much work in this field for people such as Calzada. Panama prides itself on being one of just a handful of tropical developing countries to have tamed its mosquitoes and nearly conquered malaria. American military engineers built a canal through Panama in the early 1900s, and forced malaria to retreat to the remote fringes of the country. Since then it has stagnated, primarily in its most benign incarnation, vivax malaria, which is rarely fatal.

But things have changed in recent years, and Calzada has agreed to show me some obscure signs. He emerges from the imposing Gorgas Memorial Institute, Panama's sole health research center. Clean-shaven and trim, Calzada has a slightly worried look in his eyes that is offset by high cheekbones suggesting a perpetual half-smile. I wait while he meticulously changes out of his work clothes—button-down oxford shirt and slacks—and into a T-shirt and jeans. Climbing into my diminutive white rental car and tossing a baseball cap on top of his backpack in the backseat, he patiently directs me out of the labyrinthine metropolis. Navigating Panama City's congested streets, past shiny skyscrapers and packed cafés, is a task that challenges even my well-honed Boston driving skills.

After twenty minutes heading east out of the city, the road turns quiet. It's a lovely drive, with hills in the distance, verdant pasture and scrub unbroken save for a few elaborately gated houses set far back from the road. Colombian drug lords, Calzada says, by way of explanation. Another hour passes, and the road rises, a glittering lake coming into view, just visible through a tangle of jungle. As we near the water, the pavement ends, and we pull over.

Here, at the end of the road, is the town of Chepo. From what I can see, it consists of a wooden lean-to facing a sleepy roadside café. Two police officers amble out of the lean-to, which turns out to be a checkpoint. They take my passport and vanish, leaving Calzada and me to buy a cold drink at the near-empty café. As we sit, I can just make them out in the murk within the lean-to, inspecting the blue passport with great care, turning it over and over in their hands as if for clues to some baffling mystery.

Inspection completed, Calzada leads us on foot behind the road. The hillside is green and lush, with a slick red clay track leading to the crest. He heads up and I follow gingerly.

At the top of the hill lies an improbable settlement. Packed together, not ten feet apart, are dozens of hand-built ranchos, their thatched roofs sitting on top of roughly hewn wooden poles. More arbor than hut, some of the structures rest on concrete slabs, with airy wooden-slat walls on three sides, but most are fully open-air, situated directly on the packed dirt. Inside the ranchos, smoldering fires are encircled by battered metal cooking vats, parrots sit on overturned baskets, and hammocks sway from high rafters.

From the road, Chepo seems abandoned, but in fact, three hundred of Panama's indigenous Kuna people live here, tucked away.

It starts to rain, and we duck under the eaves of a rancho. Women pass to and fro in bright puffed-sleeve cotton blouses with patterned *molas* tied around their waists and elaborate beaded anklets that reach up to their calves. They are cutting plantains, carrying plump naked children. One puts out a giant metal vat to collect the rainwater sliding off the thatch. A rooster strides by purposefully.

A half-dozen boys clad in saggy cotton underwear and wearing shell necklaces happily kick a deflated green soccer ball. One boy, around eight years old and wearing cracked red plastic flip-flops, gnaws on a green mango pit while absentmindedly pulling on his penis. A little girl walks by holding a baby covered in a rash, whom she hands to me easily. It is a tranquil scene, earthy and ripe, this hidden place at the end of the road.

It is soon apparent that most of the residents are in one of the larger ranchos, sitting around a smoky fire. Peeking in, we see them singing softly and dancing. A few are sprawled on the clay floor, facedown, passed out. We've arrived in the midst of a fiesta, Calzada whispers to me. A local girl has recently menstruated for the first time, and so the community has spent the day drinking *chicha fuerte*, a brew made from fermented corn. As we watch, a woman and a boy lift a comatose mud-caked man off the ground and drag him home. Two women from inside the rancho follow them to the doorway, smiling. Aside from a few furtive looks, they ignore us almost entirely.

It wasn't like this the year before, when Calzada first came here.

There is no English-language record of what happened to Chepo's Kuna community in 2005, save the one you are reading now. The mosquitoes that hatched from Chepo's stagnant puddles, the edges of the lake below, in the open-water cisterns, had gone on a rampage. Contaminated with the most malevolent malaria parasites known to humankind in their spittle, they alit on the exposed and unclad Kuna around them. By the time Calzada and his team arrived, nearly half of the settlement was fevered, terrified, immobilized in their hammocks.

After days of triage, Calzada brought samples of the Kuna's infected blood back to his lab at the Gorgas Institute to analyze. The most common malaria in this part of Panama is the relatively benign vivax strain, caused by malarial parasites called *Plasmodium vivax*. Instead, Calzada identified parasites called *Plasmodium falciparum*, which are more commonly found in sub-Saharan Africa. Worse, this was no regular strain of *P. falciparum*, but a particularly nasty one that had evolved resistance to standard antimalarial drugs, a trick the parasites may have picked up somewhere in Southeast Asia. Malaria experts around the world had been tracking the spread of this bug for years. At Chepo, Calzada had discovered its northernmost beachhead.

There was precious little evidence, when we first arrived in Chepo, of the village's connection to modern industrial life. In one rancho, I saw a battery-powered radio, but other than that, we might have been in the preindustrial world: there were no toilets, no running water, no electricity. But then, as the rain steadily turned the dirt lanes between the ranchos into mud, impromptu streams formed, ferrying Chepo's hidden debris to the lake: a blue plastic sandal, a crushed orange juice container, a small gas can, and a shopping bag came bobbing down the hill. We were, after all, less than two hours' drive from a boisterous city of three million, a center of international commerce through which passes 5 percent of the world's trade.[1] The scene of malaria's malevolent homecoming in this secluded settlement cast its shadow over the very doorstep of the global economy.

The 2005 epidemic at Chepo did not occur in a vacuum. On the contrary, between 1998 and 2004, malaria cases in Panama quadrupled.[2] And globally, malaria's death toll has grown inexorably since 1981.[3]

In 1995, Europe suffered ninety thousand cases of malaria. Then, in 1996, military troops in war-torn Afghanistan sparked a malaria epidemic across Central Asia. Soon, Azerbaijan, Tajikistan, and Turkey suffered malaria outbreaks.[4] By 2003, *Plasmodium* had preyed on ten times more people in Central Asia than just a decade before,[5] and a tsunami of people carrying the parasite from Africa and Asia began showing up in Europe. Today, eight times more malaria patients arrive at clinics and hospitals across Europe than did in the 1970s, and back then, most of the *Plasmodium* imported into Europe was of the vivax strain. Now, nearly 70 percent is the deadly *P. falciparum*.[6]

These days, mosquitoes infect between 250 million and 500 million people with malaria every year, and close to 1 million perish. Equally shocking is the sheer length of malaria's tenure upon us. Humans have suffered the disease for more than 500,000 years. And not only does it still plague us, but it has also become even more lethal. That's quite a feat for a disease we've known how to prevent and cure for more than a hundred years. During that same time, we've vanquished any number of similarly once-commanding pathogens, from smallpox to the plague, and have come to expect nearly complete control over newer pathogens, such as SARS or avian flu. The few that slip through our fingers, such as HIV, are the rightful subjects of anguish and soul-searching.

Yet despite the fact that we've known about malaria since ancient times, and have the drugs, killing chemicals, and know-how to avoid it, something about this disease still short-circuits our weaponry.

After dropping Calzada off, I headed back to my rental cottage along the banks of the canal, where I spent the evening reviewing my notes. The cottage was on stilts, and cooled by ceiling fans, but the window screens were old and sagging, bent and giving way from the window frame. Every morning in Panama I would awaken with some unexpected swelling from the mosquitoes' nighttime blood feasts: under my eye, on my eyelid, on the palm of my hand. Smashed mosquitoes, glued to the surface with their own internal juices, dotted the walls.

A flimsy mosquito landed gently on my forearm. A familiar spike of rage rose as I watched, incredulous, as the insect prepared to puncture my skin with her proboscis. *How dare she!* Instinctively, my hand snapped up.

Somewhere inside that cold-blooded, brittle body lurked entities whose exertions explained the making of rich and poor, sick and healthful. My hand came down a bit slower for the passing thought, and I brushed the mosquito away like a crumb. Her delicate legs snarled together, pitching the insect's body forward at a steep angle. Mangled, she skittered off my arm awkwardly as I watched, my vestigial Jain sensibilities slightly horrified. Finally she reached the precipice, where she somehow took flight and vanished.

2. BIRTH OF A KILLER

Terrie Taylor is one of the world's leading experts on pediatric malaria. Since the 1980s, she's spent six months of every year inside malaria's epicenter in central Africa, unraveling the mysteries of a disease that takes the lives of three thousand African children every day. Taylor meets me at the airport in Blantyre, a 500,000-strong city in southern Malawi, at the beginning of the 2007 rainy season. Suffering an average of 170 bites from malaria-infected mosquitoes every year,[1] between 40 and 70 percent of the entire populace of malaria-plagued nations such as Malawi harbor malaria parasites in their blood.[2] In her fifties, Taylor wears long loose skirts and keeps her frizzy brown hair parted in the middle. She starts talking straightaway, as if we've known each other for years, grabbing my shoulder and making gently irreverent cracks. She marches through the airport waving and calling out greetings to nearly everyone we pass.

The air in Blantyre, as we exit the airport, is scorching and heavy with humidity. Soon the rains will start, and the public hospital where Taylor works will be full of frightened parents proffering their limp, fevered children. During a typical malaria season, the research ward where Taylor works admits 250 malaria-infected children, of whom between 25 and 40 will die. And yet despite the passage of decades, being separated from her new husband (who is back home in Michigan), the oppressive heat, and the inevitable malaria deaths she will most certainly witness, Taylor exudes excitement. She's more like an avid camp counselor at the beginning of summer than a doctor about to minister to an epidemic.[3] She extols the friendliness of the staff at Blantyre's ramshackle airport, the beautiful views along our drive, the easy-to-clean halls of the hospital.[4] Perhaps this is her coping method, I think to myself. Or perhaps not. For Taylor is a scientist, too. In a matter of days she will venture into the beating heart of the malarial beast she's stalked for decades. As with Captain Ahab and the whale, there's a certain giddy anticipation to it.

Most pathogens mellow as they age. It's enlightened self-interest, as the theory goes. Diminishing virulence is a superior strategy for survival. It doesn't make much sense for a pathogen to rapidly destroy its victim—a dead body just means it's time to move on. Take measles and smallpox, for example. In Europe, when those pathogens first emerged, they were probably reckless killers, taking millions of lives. The survivors learned how to withstand the diseases' ravages, though, and in time both measles and smallpox settled into being unremarkable childhood illnesses, felling scores only when encountering virgin populations, such as those in the New World of the fifteenth century.[5]

Which begs the question as to malaria's tenacity and continuing malevolence. Malaria has been plaguing humans in Africa for some five hundred thousand years, with the first encounters between human, mosquito, and malaria parasite probably occurring around the time our ancestors discovered fire. Malaria existed in Africa before then, too, feeding on the birds, chimps, and monkeys that lived in the canopy.[6] We've had plenty of time—our entire evolutionary history, in fact—to adapt to malaria, and it to us. Or, at least, to devise tools and strategies to blunt its appetite. And yet, despite the millennia-long battles between us, malaria still manages to infect at least three hundred million of us—that is one out of twenty-one human beings on the planet—and kills nearly one million, year after year. As an extinguisher of human lives, write the malariologists Richard Carter and Kamini Mendis, malaria historically and to this day "has few rivals."[7] It remains essentially wild and untamed, despite its great antiquity.

And experts such as Terrie Taylor have spent lifetimes trying to figure out why.

One simple reason for malaria's ferocity is that the protozoan creature that causes the disease is, by definition, a cheater at the game of life. It is a parasite, a creature that can eke out its livelihood only by depleting others of theirs. The rest of us all do our obscure little part in the drama of life, weaving ourselves deeper into local ecology and strengthening its fabric, the bees pollinating the flowers, predators culling the herds of their weakest members. Parasites don't help anyone. They're degenerates.

Take the parasitic barnacle, *Sacculina carcini*. It is born with a head, mouth, segmented body, and legs, just like any respectable barnacle. But then, because it is a parasite, it stops developing into an independent creature. It burrows into the shells of the crabs off of which it will spend its life feeding. There it loses its segments, its legs, its tail, and even its mouth, devolving into a pulsing plantlike form, little more than a blob with tendrils sucking food from the forlorn crab's body.[8] It's the very definition of repellent. In 1883, Scottish lecturer Henry Drummond called parasitism "one of the gravest crimes of nature" and a "breach of the law of Evolution." Who can blame him?[9]

And yet parasites such as *Plasmodium* are not anomalous on this earth. According to the science writer Carl Zimmer, one third of all described species practice the parasitic lifestyle.[10] To be fair, for *Plasmodium*, parasitism arose as an accommodation to newfound opportunities, not because of any intrinsic quality or irreversible mechanism within it. *Plasmodium* did not start out life hardwired to steal. This killer first emerged on the planet as a plantlike creature, most likely some kind of aquatic algae. We know this because 10 percent of the proteins in modern-day *Plasmodium* parasites contain vestiges of the machinery of photosynthesis.[11]

Plasmodium's ancestors probably rubbed shoulders with the eggs and larvae of mosquitoes, similarly floating on sun-dappled waters.[12] When the mosquitoes took wing, malaria's ancestors likely went quietly along with them.[13] It must have happened, then and again, that when a mosquito pierced a bird or chimp or some other blood-filled creature, malaria's algae ancestors fell into the wound. Most probably died. But through the blind ticking clock of evolution, one day some subset of the interlopers found themselves thriving in those crimson seas, and a vampiric parasite was born.

Such are the ironies of surviving on this protean planet. A creature at the very bottom of the zoological scale, a humble being beneficently converting sunlight into living tissue (and thereby providing the basis for the planet's entire food chain), turns into one of the most ruthlessly successful parasites ever known, commanding two separate spheres of the living world, human and entomological.[14]

Henry Drummond would have been appalled.

Delve into even the most rudimentary scientific literature on malaria and you will soon be confronted with a dizzying range of unpronounceable words. There is *exflagellation*, *erythrocytic schizogony*, and *exo-erythrocytic schizogony*. There are *gametocytes* and *trophozoites* and *sporozoites*. These are not obscure terms for little-discussed facets of the parasite whispered over cluttered lab benches by a few old-school malaria nerds, but rather basic stages in the parasite's life cycle bandied about by nearly everyone in the malaria world, from ponytailed Harvard undergrads to queenly Cameroonian researchers and grizzled Italian vaccine makers. It is as if scientists had to come up with a whole new language just to talk about malaria.

That's because during the course of its life, *Plasmodium* transmogrifies into no fewer than seven different forms, which vary in both morphology and physiology. Its parasitic modus operandi demands such shape-shifting wiliness. After all, in order to survive, the malaria parasite must extort from two different species: the animal whose blood it feeds upon, and the insect who deposits it into that animal's blood. It's sort of like robbing a bank while stealing a car. Things get complicated.

The mosquito's immune system instinctively attacks the parasite, encapsulating the intruder in scabs and bombarding it with toxic chemicals.[15] To survive, the parasite must unleash armies of progeny in such massive numbers that fighting it off becomes more trouble than it's worth.[16] Male and female forms of the parasite, called gametocytes, then fuse, and the resulting parasites create cysts that cling to the walls of the bug's gut. (The spasmodic waving of the male gametocyte's long tail, which precedes the act of fusing with the female—yes, this microbe reproduces sexually as well as asexually—is called exflagellation.) Tens of thousands of slithering threads explode from the cysts and swarm up to the mosquito's salivary gland. This is the form the parasite must take to infect human beings. Malariologists call it the sporozoite. When the mosquito starts a blood feed, some two dozen slivery sporozoites will escape into their next host.

The parasite's shtick fails in most of the world's 3,200 species of mosquito. It works only in a single genus, called *Anopheles* (rhymes with "enough of peas"), most likely because of that mosquito's strangely tepid defenses. This restriction doesn't hinder the parasite terribly, though: there are some 430 known species of *Anopheles*, distributed in every corner of the planet except for Polynesia, east of Vanuatu. At least 70 species are known to carry malaria.[17]

Outwitting the human body's defenses, though, requires orders of magnitude more cunning. The parasite must conceal its appetite and indeed its very presence inside the body. The object of its desire—the hemoglobin inside red blood cells, which it feasts upon—is particularly precious. Produced from iron in bone marrow, hemoglobin makes it possible for blood cells to attach to oxygen molecules, and thus ferry life-giving oxygen to the body's tissues. Without hemoglobin, lone oxygen molecules maraud unattached, degrading cells, proteins, and DNA as surely as they brown sliced apples and rust metal, and the body weakens, becomes anemic, and ultimately perishes.

The parasite must hide. First, the sporozoites retreat to the liver, where they spend a few surreptitious days shifting, regenerating, dividing, and generating again, secretly transforming into an army of fifty thousand parasites in a new form capable of infecting red blood cells: the merozoite. In the next stage of the invasion, the merozoites pour into the bloodstream. They are cleverly disguised inside the liver cells they've gagged and murdered,[18] but an epic battle ensues nevertheless, and the body's immune fighters slaughter thousands. It isn't a perfect victory. If a few stragglers in this marauding horde manage to escape, they latch onto red blood cells, and within moments penetrate the cells' interior. There, they quietly feast on hemoglobin, and a new round of shifting, regenerating, dividing, and generating ensues. Some transform from tiny ring-shaped beings into fat, rounded creatures and unleash a wave of progeny. When nothing is left of the former oxygen-carrying cell besides a stream of waste and a bulge of fattened parasites, the parasites burst out of the cell and rush out to invade and consume a fresh crop of cells. Others quietly shape-shift into the male and female forms called gametocytes and lie in wait inside their hijacked blood cells. With any luck, they will be picked up by another bloodthirsty *Anopheles* mosquito.[19]

A creature this protean and multifarious defies easy challenge.

Nor is there any simple way for humans or mosquitoes to foil the parasite by avoiding the behaviors it so ably exploits.

The blood-feeding of mosquitoes, for example, is probably the most important thing a mosquito ever does—so crucial, in fact, that it risks its very life to do it. Piercing the skin of some creature many times larger than yourself is not for the fainthearted, particularly when your body can be pulverized with a simple wave of the hand or swish of the tail. Plus, blood is thick and therefore crushingly heavy for the average mosquito, which weighs significantly less than, say, a drop of water. Swollen with bloody bounty, a mosquito can barely fly, which is a mortal debility.[20]

But because velvety blood, rich with life-giving protein, is pure cream compared to the nectar they generally dine on, mosquitoes have devised clever strategies to circumvent each of these challenges. For one, they reserve blood-feeding to just a few precious moments in life, when it really counts. Emboldened by the promise of impregnation, only the female dares do it, using the rich meal to nurture her eggs. She finds her victim by following a trail of lactic acid and carbon dioxide in its exhalations. Then she numbs her chosen spot with a drop of saliva, which is spiked with compounds that deaden pain and retard clotting. Once sated with a volume of blood several times heavier than her own body, she departs immediately for the nearest vertical surface, where she spends forty-five death-defying minutes excreting all the water from her feast until, unburdened, she's once again light enough to flap her tiny wings and sail away.[21]

Malarial sporozoites that spilled into the wound with her drop of saliva, meanwhile, have by then already infected her victim's liver. But what else can the mosquito do? The survival of her progeny depends on her blood-feed.

Plasmodium's wiles similarly thwart overt human challenge, mostly because in the vast majority of cases, victims are completely oblivious to the fact of infection until it is far too late to do anything to impede the parasite's progress—even if they knew how. Almost all of *Plasmodium*'s manuevers inside the body occur in utter secrecy. When it slips into the body, while it hides in the liver, and even after it emerges into the bloodstream and attacks blood cells, there is no itch, no rash, no sweaty forehead that belies the infestation roiling within. It is only after malaria parasites rupture out of their hijacked cells, well into the parasitic invasion, that the infected person feels sick. The waste from thc parasite's hemoglobin feast leaks out of the destroyed cells, and that tiny spike of poison triggers a round of detoxification, throwing the victim into a high fever, followed by chills and shivering. When the waste disperses, the fever passes, and for several days there might be no symptoms at all—until the parasite finishes gobbling up its next batch of hemoglobin and explodes again in search of more, triggering another attack of fever and chills.[22]

The parasite's steady consumption of its victim's blood drains him of vitality, making him easy pickings for other pathogens of various ilk. But while the parasite grows inside, aside from an enlarged abdomen—the spleen of the malaria-infected can swell to twenty times its normal weight while clearing the body of dead cells[23]—its passage remains obscure. All the while, mosquitoes will bite, and imbibe the parasite roosting in the blood, and the cycle continues.

Just in case all this proves insufficient to secure malaria's safe passage, *Plasmodium* manipulates its unwitting hosts to more pliantly succumb to its will. While gestating inside the mosquito, the malaria parasite somehow manipulates the mosquito to become more cautious, seeking less of the life-giving blood it needs for its young, thus reducing the insect's risk of getting smashed or eaten and destroying the parasite developing within.[24] Once fully developed inside an infected mosquito, though, the parasite shifts its calculus, manipulating the host insect to bite more often, and more persistently,[25] by depressing the anti-clotting compound in her saliva, apyrase, so the mosquito can barely get enough. Unsated, she is more likely to seek out yet another victim, whom she can infect with yet more parasites.[26] Genetically engineering mosquitoes to resist malaria infection weakens them. Could there be something about malaria infection that helps mosquitoes stay alive, despite the parasite's selfish intentions?[27]

Not very much is known about how malaria infection manipulates human behavior to its own ends. Obviously, it leaves human victims passive, despondent, supine: in other words, more vulnerable to the bites of mosquitoes. Human attractiveness to mosquitoes depends on various factors—the smell of their feet, the chemicals on their breath, and the temperature around them—but according to studies of how mosquitoes behave around infected and uninfected people, being infected with malaria parasites alters human chemistry in some subtle way that mosquitoes find especially attractive.[28]

Certainly, malaria has left a deep fingerprint on our genomes. Today, one out of every fourteen human beings carries genetic mutations that first evolved to defend the body from malaria's onslaught.[29] This legacy has resulted in myriad conditions, reverbs of malarial destruction, some more debilitating than others.

Malaria is not just a human problem. During its long reign, *Plasmodium* has been able to spread its tentacles into a wide range of furry, fanged, and scaled bodies. More than one hundred different species of *Plasmodium* parasites specialize in infecting a veritable Noah's ark of creatures. There are malarias in chimps, gorillas, and orangutans; in thicket rats, porcupines, and flying squirrels; in pheasant and jungle fowl. Malaria parasitizes lizards and occasionally snakes. As I write this, malaria parasites teem with purpose inside the veins of the house sparrows skittering outside my window.[30]

Malaria's relentless pursuit of every available ecological niche from which it might suck sustenance includes a wide range of human habitats, from the deepest tropics of Asia to the deserts of Africa and the cool climes of northern Eurasia. In some, human-mosquito intimacy is intense; in others, distant. In some places it is sporadic; in others, continuous. At times, human defenses have been able to repel parasitic invasion; at others, we've fallen like sand castles in the tide. Throughout it all, *Plasmodium* has successfully maintained its contagion of humankind, a long pillage that has left us with no fewer than four different species of malaria parasites stalking the human race.

The first human malarias probably made a fairly marginal living. The bite of a mosquito was a relatively rare thing millions of years ago when early humans roved the savannah in search of game. If the parasite managed to get deposited into them, it probably never got out: early humans might not be bitten again for years, even decades. Stuck in dead-end hosts, the malaria parasite inside them thus would have fed for a while, reproduced, and then perished, waiting for a ride that never came. That early proto-*Plasmodium* parasite could probably have infected only about 1 percent of the thirty trillion or so red blood cells that gushed through human veins.[31]

Despite the difficult circumstances, *Plasmodium* found ways to hang on with the emergence of a species of malaria parasite called *Plasmodium malariae*, which preceded the familiar microbes that exploited the filth and population density of early farm life—measles, smallpox, cholera—by several hundred thousand years. Once it finds its way into a human body, *P. malariae* can persist in a kind of suspended animation for as long as seven decades, waiting for that fateful day when its victim might chance to be pierced by another mosquito.[32]

This was an advantage, but along with developing slowly inside the human, *malariae* also developed slowly inside the mosquito, which was a serious liability, especially given Africa's much cooler climate in pre-agrarian times. Even though they never live free in the outside environment, malaria parasites are still highly vulnerable to it, subject to the fluctuating temperature inside the cold-blooded mosquito. If the mosquito's body is warmed by summery weather, the malariae parasite can mate and produce the necessary slivery sporozoite forms in about two weeks. But in temperate weather—say, sixty-eight degrees Fahrenheit—*P. malariae*'s development inside the mosquito gets sluggish. It needs a month or more to mate and produce sporozoites. Its mosquito, by then, is long dead.[33]

P. malariae's prospects, therefore, were never that great. For thousands of years, it probably just barely hung on, perennially at risk of extinction given the uncertainty in its transmission cycle. Such is the fate of the pioneer.

Adjustments continued. Most likely there were dead ends, of which we know nothing. We do know that, in time, a new strain of malaria parasite emerged: intense, furious, awkward. *Plasmodium vivax* parasites use proteins studded along red blood cells to attach themselves to the cell before invasion. *P. vivax* could foil human defenses better than *P. malariae*, allowing it to multiply much faster within its human victims, a huge step forward from its slow-developing cousins. *P. vivax* could perform its cycle within the human host—reproducing an army of parasites, feasting on hemoglobin, and producing gametocytes, the form that is able to infect mosquitoes—in just three days.[34] So long as the mosquitoes were biting, early human communities would have found themselves at its mercy.

But *P. vivax* could not sustain a constant contagion upon early peoples, either. Mosquito carriers in human communities were still too few and far between, and the parasite still dangerously vulnerable to the cool climate of Ice Age Africa. *P. vivax*'s transmission cycle remained uncertain. For the humans, this made it much more dangerous. For when *P. vivax* did emerge, it did so suddenly and with epidemic force, burning through entire tribes like an erupting volcano, only to vanish suddenly, leaving behind piles of bodies as carrion.[35]

We know that *P. vivax* must have exerted a powerful blow because of the way early peoples responded to it, with genetic mutations that stopped the parasite dead in its tracks. The stakes for anyone with an edge against *P. vivax* must have skyrocketed. With each assault, our ancestors' immune system primed itself against the parasite, honing its ability to foil it. Its scrambling genes trotted out different permutations, holding out the possibility that one might outmaneuver the enemy.

One day a child was born who, despite seeming quite ordinary, held just such a secret weapon inside her genome. Her DNA, like that of the rest of us, comprised millions of chemicals called nucleotides, arranged in elaborate and complex patterns. One of these had been switched out of order. Such switches, rare though they are, generally result in a swift death in utero or soon after. This child was lucky. She survived despite the hidden eccentricity. The nucleotide switch resulted in her blood cells lacking a certain variety of proteins protruding from their surfaces. Functionally speaking, this didn't make a lick of difference.

As the child went about her business, mosquitoes flew into the soft wind of her exhalations, attracted by the carbon dioxide and lactic acid. *P. vivax* slid into the tiny painless wound the mosquitoes had made, and then hid in the girl's liver, as always, gathering strength for their impending invasion of the red blood cells. But try as they might, the parasites could not hold on to her subtly transformed blood cells. Without those studded proteins—they've come to be called Duffy antigens—the mighty *P. vivax* that infected the girl was rendered toothless. The foiled parasite would have floated in her bloodstream unhitched, unfed, and exposed. Patrolling foot soldiers of the child's immune system would have neutralized it with ease.

When *P. vivax* descended on the girl's band of hunter-gatherer kin, their gnarled and arthritic bodies would have writhed with fever and chills, while the girl remained healthy. She'd have extricated herself from the carnage, found a mate, and pushed out babies similarly blessed with the gift of smooth red blood cells. In time, she'd have mothered an army of vivax-immune descendants.

Stepping over the bodies of *P. vivax*'s earliest human victims, these descendants swept across the continent of Africa like a tide, reaching as far as the Arabian peninsula and the fringes of Central Asia. By about five thousand years ago, the smooth-celled woman's hegemony was complete, and there wasn't a single African alive on the continent who wasn't her descendant (or the descendant of someone similarly endowed with Duffy-less blood cells).[36]

With that, vivax malaria's reign in Africa abruptly ended.

• • •

By then, however, both *P. malariae* and *P. vivax* had long escaped their cradle in Africa. Malaria-plagued pioneers had walked out of Africa and settled across Asia, the Middle East, and Europe, ferrying *P. vivax* and *P. malariae* with them.

The temperate regions presented yet another mortal challenge for *Plasmodium*. European *Anopheles* mosquitoes were larger, more fecund, and stronger fliers than their tropical cousins. This made them better carriers of malaria. But they also hibernated all winter, to survive the continent's killing frosts. The long hiatuses between blood meals could disrupt the cycle of malaria transmission for good. *Plasmodium* can't just linger inside the body of a mosquito for months on end; it can't hibernate. After a few weeks trapped inside with no escape, it starts to disintegrate.[37]

And so vivax malaria evolved another stage in its convoluted life cycle. After entering the liver, the parasite developed the ability to transform into a dormant form called a hypnozoite, which can survive unnoticed inside the human body for months. In this state of arrested development, it waits out the winter. Later, activated by some as-yet-unknown trigger, the hypnozoite awakens, and the parasite restarts its development and its invasion of red blood cells. (The resulting attacks of malaria suffered by the victim are considered relapses, as opposed to new infections.) This adaptation allowed *P. vivax* to lie low until the bugs started biting again and its blood feasts could resume.[38]

We know that *P. vivax*'s burden must have been costly in Europe and Asia, because genetic mutations that lessened its toll emerged and spread among malaria's prey, albeit weakening those who carried them. Normally, genes that deform people's red blood cells impose enough of a disadvantage to their carriers that the gene slowly dies out, and yet in many regions of the world, such deformities persisted and spread. Hemoglobin E, a gene that deforms hemoglobin, slowed

P. vivax's progress in the body. A genetic condition called thalassemia reduced people's risk of getting sick from *P. vivax* infection. Another called ovalocytosis makes red blood cells oval and so rigid that they resist invasion by malarial parasites. Thanks to *P. vivax*, hemoglobin E spread throughout Southeast Asia, thalassemia in the Middle East and Mediterranean, and ovalocytosis through the Pacific region.[39]

But vivax malaria, in its post-Africa incarnation, was not a killer. Rather, it enslaved its victims, imposing a constant and unrelenting tax in blood. The convulsions of fever and chills arrived every summer and fall, as soon as the first mosquitoes fed on the blood of an obliviously relapsing carrier of dormant parasites. *P. vivax* infected the placentas of growing fetuses. Infected babies withered, with stunted immune defenses that rendered them vulnerable to diarrhea and pneumonia. Under the spell of chronic vivax infection, grown men and women weakened to the point that their ambitions drained away and they became anemically prone and wan, just vital enough to make more blood cells available for a later parasitic feed.

Convulsed by much more dramatic pathogens such as cholera, measles, and smallpox, malaria's victims may have barely noticed the parasite's toll.[40] The agrarian lifestyle they had come to lead—staying put on their fetid lands, weak from hunger, living together cheek by jowl—favored the spread of infectious diseases of every ilk.

Unlike the battle between *P. vivax* and the Duffy gene in Africa, which ended with *P. vivax*'s retreat, the battle between the hemoglobin-deforming genes and *P. vivax* in Europe, Asia, and the Pacific region resulted in one of *P. vivax*'s greatest victories.[41] The malaria parasite had created a new kind of mildly hobbled human, one who could withstand its invasions indefinitely.

Meanwhile, new opportunities arose in Africa.

For thousands of years after the Duffy gene beat *P. vivax* out of Africa, the continent probably carried a fairly light burden from malaria. Its nomadic tribes would have encountered malaria-carrying mosquitoes only occasionally. Common ones, such as *Anopheles arabienses*, lived on Africa's dry savannahs, and fed mostly on animals, not humans. The *Plasmodium malariae* parasite hung on, but just barely, as did an even rarer human malaria parasite, *Plasmodium ovale*.

But with the formation of the Sahara desert about twenty-five hundred years ago and the spread of Bantu-speaking peoples into the equatorial rain forests of the continent, humankind and mosquitoes collided together in novel ways. The Bantu hacked into the rain forests to grow yams and plaintains, thus transforming those areas. As the trees fell, the chimps and birds disappeared. For the first time, shafts of sunlight reached the rain-forest floor and it became denuded of the thick absorbent layer of humus that once blanketed it. Rainwater collected in puddles on its rutted surface.[42]

A new species of *Anopheles* mosquito emerged to exploit these new ecological conditions, one that constantly, ingeniously, tests out new habits and habitats. Today, of the nearly 500 known species of the *Anopheles* mosquito, 170 belong to so-called species complexes—mosquitoes that have, for mostly unknown reasons and despite sharing the same habitats, stopped mating with one another. They're tribes of the same species, living on the same land, who have banned intertribal romance. It's an absurdity on the face of it, given that mosquitoes by and large live solitary lives. They're not sociable enough to be so purposely antisocial. And yet, species complexes have formed again and again.

So it must have been during the early days of the first farm villages in the rain forests of Africa. A tribe of *Anopheles* colonized the rain-forest villages, laying their eggs in the new, sunny puddles, which were conveniently as free of fishy, larvae-eating predators as the tree holes of the canopy. These mosquitoes developed a taste for the plentiful humans, within easy reach. In time, this mosquito tribe begat a species: *Anopheles gambiae*. (The malariologists I spoke to pronounced *gambiae* to rhyme with *Bambi*.) *A. gambiae* enjoy the good life.[43] Specifically adapted to the world in and around human habitations, they don't even need a strong proboscis or swift flying skills.

Unlike anopheline species in other parts of the world, which feed on humans less than half of the time, *A. gambiae* specialize in humans, rarely if ever extracting a blood meal from anything other than *Homo sapiens.*[44] There would have been very few animals to tempt *A. gambiae* in these rain-forest villages anyway. The yam and plaintain farmers didn't keep domesticated animals, as their counterparts in Europe did. For one thing, African mammals were notoriously difficult to tame. For another, the local tsetse fly transmitted deadly sleeping sickness to any cow or horse brought into Central Africa.[45]

All of this held great import for the malaria parasite. In almost every other place where it lived, it had to contend with the imperfect carriage of its mosquito vectors. It had to contend with *Anopheles* species that were fickle in their blood tastes, depositing the human malaria parasite in cows or horses or some other inappropriate creature. It had to contend with mosquitoes that hibernated all winter or died out during inclement weather. It had to contend with mosquitoes that bit only sporadically, thanks to wandering nomads. All of those frustrations were relieved with the arrival of *Anopheles gambiae*, which would become the most reliable and efficient vector the parasite ever had.

All *Plasmodium* had to do to harvest the promise of *A. gambiae* carriage was subvert the Duffy defense of the locals. Any mutant parasite that could do this would enjoy the most abundant blood feasts ever, the parasitic easy life with no need to suffer through a killing winter, uncertain transport, or annoyingly wandering nomads. Under such conditions, its progeny would sweep through the population.

With sufficient numbers of malaria parasites hanging on in the margins of the continent, such a mutant parasite did one day emerge, tailored to exploit the rich new conditions. This parasite didn't rely on a single strategy to attack red blood cells, as did *P. vivax*. It had multiple invasion strategies, ensuring that its progeny would indeed get their hemoglobin. The mutant begat a tribe that begat a species:

Plasmodium falciparum (pronounced fal-SIP-ah-rum), which could infect as much as 80 percent of its victim's blood, some twenty-four trillion cells, forty times more than its cousin *P. vivax* could.[46]

The *P. falciparum* parasite doesn't need to select the weakest red blood cells, as *P. vivax* and *P. malariae* do, which prey primarily on young and aging cells, respectively.[47] It has novel ways of eluding its victims' immune system. Once inside a red blood cell, it sends out its progeny clothed in not just one but a multiplicity of disguises. Each contains multiple copies of the genes that control its antigens, the proteins that alert the immune system to launch an attack. The progeny can control the variable expression of these genes, so that when they bombard the body, each cloned parasite looks different to the immune system. Some portion may be recognized and destroyed, but those in the most novel disguises are not, and with the immune system fooled, they are able to burrow into blood cells, victorious.[48]

After thousands of years of malarial respite, *Plasmodium* was back, in its most terrifying, ferocious incarnation of all.

Plasmodium falciparum's toll on early rain-forest villages would have been devastating, with both adults and children felled by the dozens, their parasite-infested blood curdled in their veins. We know of the parasite's bloodshed some four thousand years ago because of the calamitous genetic adaptations that arose and spread.

A single-point mutation on a single gene turns pliable blood cells into rigid, frozen crescents called sickle cells. A newborn endowed with two copies of this sickle-cell gene, before the advent of modern medicine, would not have survived childhood. But those endowed with a single dose of the sickle-cell gene—the so-called heterozygotes—found themselves with a useful weapon against the scourge that stalked them. When *P. falciparum* invaded their bloodstreams and started taking down oxygen-sucking hemoglobin, the rising level of unhitched oxygen would trigger a switch. The hemoglobin molecules in their blood cells fused together like two magnets, turning into stiff crescents, and slashing the risk of death from *P. falciparum* by 90 percent.[49]

If two such sickle-cell gene carriers started a family together, the probability of their children being born with the uniformly fatal condition of a double dose of the sickle-cell gene ran to one in four. Even with those terrible odds, the carriers of the sickle-cell gene outreproduced those without the gene. *Plasmodium falciparum* was so deadly that it was better to risk a 25 percent probability of a dead child than to forsake the possibility of a weapon against the parasite.[50] And so the sickle-cell gene spread throughout the five continents, lurking inside up to 40 percent of the population in parts of Africa, South Asia, and the Middle East to this day, a silent reminder of falciparum malaria's deadly legacy and the mortal risks that surviving it has required.[51]

Humankind devised a few other weapons against *P. falciparum*. With each bout of fever, our immune systems can arduously prime themselves against yet another of the parasite's multifarious disguises. The more infections we suffer, the savvier to *P. falciparum*'s antigenic guises our immune systems become, which allows people with multiple exposures to enjoy a modicum of immunity to the parasite. They still get infected, but with their immune systems restraining the parasites' numbers to as much as one million times lower than in those not immune, they may not get as sick. They will almost certainly not die.[52]

But such immunity occurs only when people are exposed to chronic infection—that is, multiple death-defying encounters with *P. falciparum*. It is as fleeting as a suntan. After all, malaria parasites reproduce new generations at a rate two hundred times greater than we do. A few months' respite from the fiery glare of *P. falciparum*, and whatever immunity to the local parasite had been arduously acquired starts to fade away, canceled out by a new generation of the parasite.[53]

As a result, every year, many hundreds of thousands of people must face an invasion by falciparum malaria utterly defenseless. Through happenstance or lack of time, they have not built up any acquired immunity. *P. falciparum* destroys nearly one million of them every year. First and foremost, it kills the babies.

Blantyre's Queen Elizabeth Hospital is a sprawling, dusty complex of squat brick buildings surrounded by loud, traffic-clogged roads. People mill across its grounds, mostly women wearing traditional wraps and silty blouses, carrying swaddled children on their backs, hanging laundry out to dry along the gates. Inside, the hallways are packed with patients clutching tattered health booklets, waiting to be seen by a doctor. The painted turquoise wall is affixed with an orange plastic laundry basket filled with garbage, and a small sign that reads "Osalabvula." *Do not spit.* There are no fans, no air-conditioners, and although the unscreened, glass-paned windows are flung open, the air is fetid and still. Roaming down the hallway, one passes through several mild but discernible zones of odor: mold, sweat, urine. Inside the crowded wards a handful of white-coated doctors drag their stools across the floor and lean in to each patient to hear their whispered complaints.

The pediatric research ward, at the very edge of the hospital complex, consists of two large rooms holding about fifteen wooden raised beds each, a narrow, fluorescent-tube-lit hallway, and some barren, closet-size offices, including Terrie Taylor's. Unlike the rest of the hospital, with its crowds and smells, the research ward has a certain serenity to it, despite the drumbeat of child deaths that occur within its walls. Most of the young patients here are deathly ill with malaria, and comatose. There is no welter of plastic tubing or beeping machines around them, as one would see in the West. Their small bodies rest on the high beds unadorned. They appear to be simply asleep.

Under their beds, their mothers and grandmothers have unrolled their thin wraps and are resting on the cool cement floor. When Taylor and her team stride into the ward, the women jump up abruptly, like schoolkids who've been sneaking a nap.

Two-year-old Duke arrived at the hospital on the Friday before I came to Malawi. He'd been visiting relatives living near Blantyre when he suddenly fell terrifyingly ill. His mother and aunt—his father was at home in their village, two hours north of the city—brought him to Queen Elizabeth Hospital. Duke most likely shared one of the steel-frame beds with another patient, while his family camped out on the hospital grounds so they could bathe him and wash his bedding. They'd have joined the hordes of others forsaking the demands of an unforgiving corn crop back home—Malawi is a nation of subsistence farmers—to provide hospitalized relatives with this basic nursing care. There's no one else to do it.

After Duke's breathing grew labored, he fell into a coma. His muscles started to flex and extend involuntarily into stiff, bizarre positions. It was in this state that Taylor's team discovered him and rushed him to the research ward.

While he lay unconscious, they pumped him with anticonvulsants and drew his blood, dabbed a ruby drop on a glass slide. Down the hall, in the sole air-conditioned room in the ward, a lab technician focused a microscope on the slide and spied the lavender, pale-centered spheres of his red blood cells. For each healthy one, there was another that had been invaded and occupied by *Plasmodium falciparum*.

Ironically, while the symptoms of severe malaria are alarmingly apparent to the victim's family, the disease isn't easy for clinicians to diagnose. The only surefire way Taylor's team can finger severe malaria is by lifting their comatose charges' eyelids and spending up to thirty minutes examining the backs of their eyeballs. A normal eyeball is gray and laced with a thin spiderweb of red blood vessels. In a patient in the throes of severe malaria, those vessels are burst, leaving the eyeball speckled with white splotches and red spots. The spidery vessels themselves are pale orange, not red, the parasite having eaten all the hemoglobin.[54]

Duke and the other comatose children who appear in Taylor's ward are surely sick, and they are often clearly carrying the malaria parasite, but this isn't enough for doctors to know that they are sick *with* malaria. The typical Malawian kid might suffer a dozen episodes of malaria within the first two years of life and think nothing of it. Which is why if you asked a local if he'd ever had malaria, as I foolishly once did, his reactions would be similar to how a New Yorker or Londoner might respond if a reporter asked if he'd ever had a cold. "Yeeessss," he'd say, eyebrows raised. As in: "And what of it, pray tell?"

Thanks to their acquired immunity, only a fraction of Malawians with the malaria parasite living in their blood get seriously sick from it, suffering the fever and chills of "clinical" malaria. Most of them are the "uncomplicated" cases, which resolve themselves on their own. But sometimes, out of nowhere, an otherwise routine falciparum infection will turn vicious. After suffering through the usual few days of fever, instead of slowly recovering, the victim will start convulsing and fall into a coma. Older victims suffer kidney failure, and their lungs collapse or fill with fluid. If they are not treated rapidly, they will almost surely die. Even if they are treated with the very best and most effective therapies, 20 percent will die. Nobody knows why.

Some contributory mechanisms are understood, at least in outline. Blood cells infected with falciparum parasites became sticky and get clogged in the small blood vessels. This is likely a survival tactic for the parasite. Stuck inside the vessels, it avoids being washed into the bloodstream and neutralized by the pathogen-killing spleen. There it can grow and develop undisturbed.[55] But the clogged vessels deprive the victim's tissues of oxygen, and clogged microvessels in the brain can starve the brain and bring on coma.[56]

Malarial coma doesn't always result in death; four times out of five the comatose patient wakes up, gets out of bed, and is able to walk home (albeit with subtle cognitive deficits, for some). Perhaps such patients are blessed with some as-yet-unknown genetic endowment. Another theory is that those who die are infected by some especially virulent tribe of falciparum parasite. Often malaria patients are infected by several different, genetically distinct strains of *P. falciparum*. Taylor has seen, under the microscrope, three different parasites gorging upon the same blood cell. Perhaps the winning parasite in such struggles is the most aggressive one, which somehow triggers the kind of malaria that kills. Or perhaps mortality in these cases is a result of the multiple infection itself.[57] But then again, says Taylor, there's no real evidence that the distinctions observed between parasite strains have any significance at all. They could be as meaningless as different-size spots on a dog.

What we do know is this: this rare and precipitous complication of falciparum infection—called simply severe malaria—develops in perhaps 2 percent of all clinical malaria cases[58] and accounts for nearly 90 percent of all the deaths from malaria worldwide.[59]

On Duke's second day on the ward, the antimalarial drugs and blood transfusion his doctors gave him started to work. The drugs starved the parasite in his cells, and its numbers started to fall. His blood thickened with healthy red blood cells. He stopped posturing and convulsing, and his fever dropped. His gaze steadied. Taylor and her team were thrilled, jubilant.

By the afternoon, even as the population of parasites in Duke's body continued to ebb, the speckling of his eyeballs multiplied. That evening, suddenly, Duke stopped breathing. Taylor's team rushed to his side. His heart thrummed an even, loud rhythm. But one by one his organs failed. His heartbeat kept going. "So strong!" remembers Taylor. "You just can't stop trying!" They tried to resuscitate him for two hours.

A few days later, I stand in the clinic's lab examining the boy's cauliflowered brain, which Taylor has had excised and pickled in a small Tupperware container after the boy died. Since 1996, Taylor and her colleagues have been conducting an extraordinary study in which the bodies of children who died of malaria are autopsied and scrutinized for clues to the series of events that led to their deaths.

Duke's brain is riddled with tiny spots, signs of some epic battle with the falciparum parasite. "Is that what killed him?" I ask. Taylor isn't sure, even holding the boy's brain in her hands. "I have no idea," she says softly, swishing the container.

Before I leave, Taylor insists I see the miscreant itself. "Show her how pretty the parasites are in red cells!" she instructs the lab tech. Peering into a microscope focused on a drop of infected blood, falciparum parasites are clearly visible. For Taylor it is an image fraught with mystery and power. "They're beautiful!" she blurts.

I go to find my bicycle, locked outside. The rain has started, and as I hurry through the halls, earthy odors rise through the moist air. Soon the rain is pounding on the corrugated tin roof, and I must wait. An orange dog similarly takes cover under the eaves in the gulley, trying not to get wet.

There's a tiny room in the hospital, at the end of a hall filled with desultory waiting patients, where the facility's medical records are kept. Despite the caprice of the pathogens that stalk its patients, the enduring mystery of how they kill, each piece of paper describing the carnage finds its way there and is collected into eight-inch piles, carefully tied with string. Part of Duke's story, like that of thousands of others like him, certainly rests in those pages, a final whisper of his short life after his distraught family trekked back home to their village. Researchers such as Terrie Taylor extract bits into computerized databases in hopes of one day deciphering those yellowing stacks. For now, they line the walls of the tiny room from floor to ceiling. The rain ends as abruptly as it started. The door is ajar as I pass by on my way to the exit, and I glimpse the shelves, buckling under the weight.

3. SWEPT IN MALARIA'S CURRENT

When historians consider how infectious diseases have shaped human history, they don't generally make much of malaria. Other pathogens—smallpox, measles, plague—have made a much bigger splash, by piling up the bodies, fast. But malaria's tide also has sculpted our landscape, in more subtle but equally long-lasting ways, from refashioning the ethnic makeup of sub-Saharan Africa to determining settlement patterns in the Americas and even influencing the creation of Great Britain.

There was a strange moment at the Multilateral Initiative on Malaria's international scientific meeting held in Yaoundé, Cameroon, in 2005. Malariologists from Europe and North America and a sprinkling of African scientists had descended upon the sweltering city to deliver earnest presentations on the latest techniques in the fight against malaria.

One afternoon early on in the conference, a local television journalist took the stage; how he'd gotten on the program is unclear. In a tailored suit, he had far outdressed the scruffy scientists assembled in the audience. His address was brief—which was probably for the best, as he quickly overturned the central premise of the meeting. He began by recounting his own malarial fevers, but not in the hushed stentorian tones of the others. For him, malaria was more like the annoying cousin who comes to visit over the holidays. And then he implied that malaria wasn't Africa's scourge, and that the malariologists present weren't Africa's saviors. Au contraire: malaria, he said, had *saved* Africa; it was the scientists who were the invaders.

Channeling the nineteenth-century king of Madagascar who famously boasted that no invader could take on his country's *hazo* (impenetrable forests) or *tazo* (its malignant malarial fevers), he extolled "General *Anopheles*" for thwarting the armies that would have attacked Africa.[1] And now, "you scientists," the journalist said, smiling and gesturing at the crowd, "are trying to take him on again!"[2]

His speech, with its somewhat ungracious equation of malaria scientists with imperialist intruders, passed without comment, and the conference continued as before, focusing on malaria's undeniable burden upon its local victims. There were slide shows. There were tours of malaria clinics. There were pleas for more funding, for better statistics, for more public attention. But through it all resonated the message behind the journalist's words: Not everyone living under *P. falciparum*'s spell considers their situation an unmitigated misfortune. It isn't easy to live with malaria, but those who survive the gauntlet of a falciparum-infested childhood gain a powerful immunological advantage over others. *P. falciparum*, deadly for outsiders, can no longer kill them.

The Bantu-speaking farmers who spread from western Africa and throughout the interior of southern Africa were among the first peoples whose relative immunity to *P. falciparum* likely vanquished potential rivals. Linguistic evidence shows that the three hundred or so different language groups found in the wide expanse of southern Africa diverged from a shared proto-Bantu language three thousand to four thousand years ago. There was, at one time, a greater diversity of linguistic groups on the continent. Remnants of these hunter-gatherer populations remain, speaking non-Bantu-derived languages, living mostly on more marginal lands, and often in positions of political subordination.

Somehow, over a few thousand years, the Bantu-speaking farmers spread across the entire continent, pushing nomadic hunter-gatherers to the edges. How did they achieve this rapid hegemony? Some say the Bantu enjoyed better diets or were more fearsome warriors. But neither of these theories fully accounts for their swift dominance.

From what we know about the early Bantu peoples, they would have had far greater tolerance to falciparum malaria than the hunter-gatherers. Here's why: Living in fixed agricultural communities where *Anopheles gambiae* roosted would have subjected the Bantu to year-round falciparum infection. Endemic falciparum malaria would have regularly killed their babies, children, and first-time mothers. But those who survived pregnancy and infancy would have enjoyed an advantage that no outsider could have claimed. *P. falciparum*—at least the local strain to which their immune systems had been primed—could no longer kill them. It would have weakened them and made them sick on occasion, but no adult from an endemic region would have died from falciparum malaria.

Nomadic hunter-gatherers did not create or live around *A. gambiae* habitat as the Bantu did. Wandering across dry areas, where *A. gambiae* were scarce, meant that their exposure to *P. falciparum* would have been minimal. On occasion, and more often as the malarious farmers multipled and spread, these falciparum-naïve bands of hunter-gatherers would have happened upon an infested settlement, and *P. falciparum* would have rampaged unfettered in their virgin bodies.

Over time, the accumulation of such encounters would have decimated once wide-ranging African communities such as the Khoisan, Pygmy, Cushite, and those speaking Mande and Atlantic languages, and pushed their bedraggled survivors to the periphery of the continent, where they remain to this day. The immunological fence that *P. falciparum* built around the Bantu repelled incursions by outsiders as effectively as a standing army. The Bantu villagers didn't have to be bigger or stronger to beat back the nomads: a couple of bites from their mosquitoes did the trick.[3]

• • •

Diminished to the point of vanishing in Africa some ten thousand years ago, *P. vivax* reemerged some five thousand years later in ancient Egypt, Greece, India, and China. Antigens to *Plasmodium* in five-thousand-year-old Egyptian and Nubian mummies and references to malaria in four-thousand-year-old Sumerian and Egyptian texts testify to the parasite's arrival.[4] Also, the ancients described the disease in vivid enough terms for historians to confirm its identity. Ancient Greeks understood malaria as a seasonal scourge that arrived during harvest time. The physician Hippocrates described it as a disease common around swamps, while the poet Homer referred to malaria when he decried Sirius as an "evil star" that was the "harbinger of fevers."[5] The ancient Chinese called malaria the "mother of fevers," while in India thirty-five hundred years ago it became known as the "king of diseases,"[6] personified by the fever demon Takman. The Vedic sages accurately described malaria's signature chills and fever. "To the cold Takman," they wrote, "to the shaking one, and to the deliriously hot, the glowing, do I render homage. To him that returns on the morrow, to him that returns for two successive days, to the Takman that returns on the third day, shall homage be."[7]

By the time of Christ, *P. vivax* had swept over temperate Europe, and it entered northern Europe during the early Middle Ages. By the sixteenth century, *P. vivax* was deeply ensconced in Europe and Asia, *P. falciparum* held Africa in its thrall, and the Americas teemed with mosquitoes with pristine, parasite-free guts. As human populations and the *Plasmodium* parasites in their veins collided during the age of exploration and conquest, malaria's differential killing power shuddered through the continents, altering the fate of nations.

On the continent of Africa, the falciparum malaria to which locals had become partially immune violently repelled unexposed European explorers and would-be conquerors. When European armies attempted to capture African gold mines in the 1570s, *P. falciparum* felled their soldiers, sleeping sickness took down their horses, and the Africans' iron weapons vanquished the rest.[8] *P. falciparum* killed the majority of the party of Portuguese missionaries and soldiers sent up the Zambezi River in 1569; in 1841, it sickened 80 percent of Thomas Fowell Buxton's party of 159, sailing up the Niger.[9] Fevers killed 88 out of 108 Europeans on an 1825 expedition into the Gambia.[10]

For centuries, Europeans could sustain little more than lightly manned trading posts in Africa.[11] Malaria and other African diseases to which the Europeans had no immunity usually killed half of any group within a year of their arrival on the continent. That mortality rate, writes historian Philip Curtin, was "simply too high" for more intensive occupation.[12] The British wouldn't even risk exiling their convicts there. After the loss of its American colonies, British Parliament debated the notion of sending convicts to Gambia, but noting that such an exile amounted to a death sentence, they decided against it. They sent them to the terra incognita of New South Wales, Australia, instead.[13]

Most scientists agree that the Americas were malaria-free for thousands of years before Europeans arrived. The first human settlers may have had malaria parasites in their blood and livers when they crossed the Bering Strait eleven thousand years ago, but the parasites would have withered and died during the slow, frigid, mosquito-free journey.

When colonists from England arrived in the early 1600s, they had parasites roosting in their veins, and they encountered a land teeming with mosquitoes and wetlands, much more so than today. Swamp, bog, wetland, and marsh covered more than 220 million acres of the region that would become the United States.[14] A beaver population roughly forty times that of today's flooded even more land every year.[15] And this temperate mosquito wonderland was connected by a narrow isthmus to the tropical South American continent, which was dripping with dense mosquito-friendly rain forest.

The New World's mosquitoes included an array of *Anopheles* species. *Anopheles crucians* spread its wings in the Floridian cypress swamps. Clouds of *Anopheles punctipennis* swarmed the continent's great temperate forests, and *Anopheles quadrimaculatus*[16] gathered at the edges of sun-dappled lakes. In Europe, malaria parasites had rendered such watery environs uninhabitable, but Native Americans were able to exploit the lush and abundant wetlands to the fullest.

In the spring of 1607, the English sailed gingerly into the Chesapeake Bay and up the James River. Wary of attack from the local Algonquins—as well as from the Spanish, farther south—they denied themselves the dry, fertile ground farther inland in favor of a small island in the river, separated from the mainland by a narrow channel. Whatever they gained in strategic security, though, the Jamestown colonists lost in vulnerability to the local *Anopheles*, for marshes covered the low-lying island, and the mosquitoes that hatched from them likely started biting the colonists not long after their arrival.[17]

The English called vivax malaria the ague (the word rhymes with "plague you"), and we know that at least one of the Jamestown settlers was probably infected with it when he arrived. Shortly before his departure for Jamestown, Nathaniel Powell described his malarial state in a letter: "I have not yet lost my quartane Ague," he wrote, "but as I had him yesterday so I expect him on Thursday next."[18] Given the relatively quick transit time across the Atlantic (compared, at least, with the earlier traverse of the Bering land bridge), infected colonists such as Powell almost certainly ferried viable vivax parasites to Jamestown.

Malaria's arrival wouldn't have emerged as a major force to be reckoned with at first. Measles and smallpox destroyed the local native peoples with cruel efficiency, while diseases of filth—typhoid, dysentery—ransacked the Jamestown settlers.[19] But those pathogens don't have much staying power. Measles and smallpox burned through all the virgin blood available to them and then receded to near oblivion. And as soon as conditions in new settlements improved even marginally, dysentery and typhoid started to recede, too. What the settlers had to live with, in the long term, was malaria, cresting like a shark on the undertow.

By the mid-1600s, the surviving Virginia settlers took over the local Native Americans' drier, more fertile lands, but in an age of water travel they could never stray too far from the mosquito-ridden lowlands by the rivers and bays, and by the late 1600s, vivax malaria had become endemic. Living in Virginia required suffering what residents called a "seasoning"—that is, "two or three small fits of a feaver and ague," as one settler wrote in 1687.[20]

The penalty of "seasoning" and the subsequent debility of chronic malaria infection reduced Jamestown and the other Chesapeake colonies to a sorry state. Chroniclers of the day described the survivors' dirt-floored, windowless houses riven with cracks, and their farms littered with rotting stumps.[21] They'd be lucky to pull in three pounds sterling in annual profit.[22] Those who did prosper—such as Powell, who joined the governor's council in 1619—enjoyed a modicum of immunity to *P. vivax*. Like Powell, many had grown up in low-lying Kent and Essex, two of the most malarious counties in England.[23]

While the Chesapeake colonies suffered the consequences of their malarial contamination, the absence of the disease leveraged the unlikely success of those European settlers who headed farther north. New England's generally cool weather slowed the development of the malarial parasite so that its life cycle far exceeded the average life span of an infected mosquito. (At sixty-five degrees Fahrenheit, *P. vivax*'s reproduction inside the mosquito slows to a twenty-day cycle, and *P. falciparum*'s to twenty-three days, while the average anopheline might live just over a week.) In addition, the rocky creeks and forested hills of New England provided few suitable habitats for the most efficient malarial mosquitoes.[24]

The Pilgrims and Puritans who settled the northern colonies knew the colder climate would be healthier. "A sup of New England's aire is better than a whole draught of Old England's ale," wrote one New England colonist in 1629. "Experience doth manifest that there is hardly a more healthfull place to be found in the world that agreeth better with our English bodyes."[25]

Indeed, although they suffered the hostility of the native peoples and the pioneer's diseases of filth like the southern colonists, and had to put up with New England's marginal growing conditions and brutal winters to boot, the New England colonists experienced a boom, right from the beginning. Within decades of settlement, the relatively small numbers of settlers there had created colonies as large as those in the Chesapeake Bay. By 1700, some twenty thousand settlers in New England, through birth and minimal immigration, grew colonies as big as those in Virginia and Maryland, where more than 140,000 migrants had landed. The average life expectancy hovered around sixty years, with each new generation doubling the size of the population.[26] The superior disease environment of the North—in particular, its relative freedom from malaria—spelled the difference.

Oceans of ink have been spilled on the cruelty and waste of the so-called triangular trade that European colonists established in the Americas over the following centuries. For the love of sugar, easily grown in the tropical Americas, and silver, mined from its lush deposits, millions of Africans were shackled and enslaved and shipped across the sea, and the products they created were looted and pirated across the sea to sate Europe. That bloody trade, which cost millions of lives and disrupted countless cultures and economies, spread over four continents, and we tremble from its reverberations to this day. Malaria did not create it. But *Plasmodium* certainly helped carve its harsh contours.

By the mid-sixteenth century, labor had become a problem in the New World colonies. The local native peoples, enslaved to work the sugarcane crop and the silver and gold mines, were dangerously vulnerable to easily transmissible Old World germs such as smallpox and measles, broadcast on the Europeans' breath, coughs, sneezes, and dirty blankets. Wave after wave of disease—a single smallpox epidemic during the 1560s, for example, felled thirty thousand native slaves in Brazil[27]—soon denuded the cane fields of workers.[28]

It might have seemed reasonable, at this point, to transplant American sugarcane colonies to West Africa, which was similarly amenable climatically and much closer to European markets. But falciparum malaria had already rendered such arrangements untenable.[29] Instead, shorn of Native American slave labor, European colonizers in the New World turned to slave labor from Africa. A small Portuguese trade in African slaves had started in the 1400s, selling African slaves captured in wars to other African states. First the Spanish and Portuguese, but soon the English, French, and Dutch as well, increasingly turned to slaves to sate their labor needs in the American colonies.[30]

But by the eighteenth century, Africa's trickle of war captives and outcasts would no longer suffice to service the increased demand for slaves. Soon raiders started plunging deeper into the interior of the continent to capture the terrified sons and daughters of peaceful farmers and villagers, to forcibly sell them into slavery. For fear of the raiders, whole villages in Africa abandoned their lands and long-established trade routes, and the exquisite balance between man and parasite forged over millennia was abruptly ruptured.[31]

The new labor arrangement proved costly for the Europeans as well. Ferrying slaves from Africa to the West Indies, and the West Indies sugar they produced back to Europe, involved a tremendous amount of difficult and time-consuming shipping—and thus exposure to disease. Slave raiders could avoid contact with foreign strains of *P. falciparum* and other diseases by passing their captives along a chain of traders, from the interior to the waiting ships on the coast. But the crews aboard the slave ships were dangerously vulnerable.[32]

The burden of disease decimated state-backed efforts by the Dutch, English, and French to monopolize the slave trade. Instead, ad hoc bands of merchants funded perilous voyages from Europe to Africa to the West Indies and back, coercing their crews into the job and not expecting to make more than one death-defying trip in a lifetime. As a result, crews aboard the slave ships usually had to wait for months at slave ports while their holds filled with screaming captives, who trickled in at a rate of two to three a day. Forty-five percent of European sailors on slave ships died, more than in any other trade of the era,[33] from the "noxious vapour, arising from the swamps," as a slave-ship surgeon of the time called it. "This trade," he wrote, "may justly be denominated the grave of seamen."[34]

Despite these conditions, between 1700 and 1800, European slavers brought some six million bound and shackled Africans to the Americas. With falciparum parasites roosting in their veins, they changed the face of the Americas forever.[35]

While malaria-experienced Europeans knew of the dangers of malarial fever, and the particular deadliness of Africa's fevers, they had no way to predict what would happen upon flooding the Americas with falciparum-ridden Africans.

Sixteenth-century Westerners didn't have a specific definition for the disease of malaria. They didn't know about the parasite or the mosquito that carried it, and so their experience of malaria led them to consider it primarily a disease embedded in the landscape. Western medical authorities of the time attributed many diseases, including malaria, to smelly vapors they called miasmas, which purportedly rose from stagnant water, putrefying vegetation, and animal remains. According to miasmatic theory, miasmas grew more dangerous in warm climates, where everything got stinkier faster, whether it was spoiling food or decomposing vegetation. This was why, as Plymouth colony leader William Bradford wrote, "Hott countries are subject to grievous diseases."[36] The diseases of the hot climates frightened sixteenth-century Europeans so much they thought that sudden exposure to heat could literally melt the fat inside a person.[37]

The potency of miasmatic theory, which held sway from the Middle Ages until the end of the nineteenth century, may have derived partly from how well its teachings helped explain and prevent malaria. Before the nineteenth century, when miasmatic theory inspired sanitationists to separate food and water from waste, vanquishing many infectious diseases in the process, the teachings of miasmatism were relatively useless. Miasmatism's admonition to avoid stagnant waters and stinky vapors did little to help people stay away from unwashed hands and food, which is what gave them dysentery and typhoid. Nor did it encourage them to avoid exposure to the exhalations of the sick, which gave them measles, smallpox, tuberculosis, influenza, and pneumonic plague. Miasmatic theory didn't help much with leprosy, either, which people got from constant, long-term contact with sufferers, or with typhus, which they contracted from lice, or syphilis, which is transmitted through sexual contact.

It did help protect people from malaria, however. The miasmatic theorists may not have known it, but the larvae of many *Anopheles* mosquito species live in stagnant waters, feeding on and hiding from predators under the rotting vegetation that endows swamps and wetlands with their distinctive sulphurous odor. Minimizing exposure to this malodorous air—by avoiding swamps and closing doors and windows, as miasmatism suggested—certainly would have helped people avoid mosquito bites. Indeed, according to miasmatic theory, it wasn't the miasma itself that was poisonous, but rather the "miasmata," or bits of debris, it carried within it. Miasmatism's prediction that warm climates threatened greater disease was especially true for malaria, for higher temperatures allowed the parasite to develop faster, ensuring that more infected mosquitoes would transmit the disease.

Miasmatic theory explained the deadliness of Africa's tropical fevers, and the relative healthfulness of cool highland areas relative to marshy lowlands. But what it couldn't foresee were the malarial parasites munching on the hemoglobin of enslaved Africans, and what would happen when American *Anopheles* took their first sips of the Africans' infected blood.

By the 1500s, the crucial land bridge between the Pacific and the Atlantic—the isthmus that would become Panama—had been rendered a noxious "land of pestilence," as medical historian James Simmons puts it.[38] The Spanish used the isthmus to transport the fabulous riches of their American colonies back to Spain. Having slaves trek the loot on foot across Panama, where the distance between the Pacific and the Caribbean spanned just forty miles, saved weeks of shipping around the tip of South America.

But those forty miles wended through one of the hottest and wettest places on the planet, covered with dense rain forest. A spine of steep mountains towering over seven thousand feet high sliced through its middle.[39] Rain pelted down in sheets for three quarters of the year, after which the winds arrived, uprooting trees and turning the steaming jungles into a vine-covered matrix of still, green pools.[40] Along the coasts, mangrove trees plunged their spiderlike roots into watery sands, creating sprawling coastal swamps, a dark netherworld of neither sea nor land. The difficult and exposed journey would have ensured steady contact between the African slaves' parasites and Panama's mosquitoes.

Once the American *Anopheles* started transmitting falciparum parasites, the Spanish found themselves utterly defenseless. By 1584, the fevers at Nombre de Diós, the Spanish village on the Caribbean side of the isthmus, were so bad that the king ordered it abandoned altogether. The new village they established, Portobelo, was not much better, and the Spanish soon knew it, too, as a "breeding place of malignant fevers," according to Simmons.[41] Portobelo was an "unhealthy place," one visitor wrote in 1648, "very hot, and subject to breed Feavers, nay death." The famous navigator-cum-pirate Sir Francis Drake died of fever not far from there, and was buried in a lead coffin under the soft Portobelo soil.[42]

In 1534, the Spanish had surveyed Panama in hopes of building a canal, but after contaminating the isthmus with falciparum malaria, they were forced to abandon Panama's fevered jungles to the indigenous Kuna people and the bands of escaped slaves the Kuna sheltered.[43] At the very height of its powers, the Spanish Empire managed to build just a single muddy mule track through the Panamanian jungle and two fever-racked villages at either end. Under constant attack by the pirates who prowled the Caribbean and by falciparum malaria and yellow fever on the isthmus, Panama became the notorious Achilles' heel of the Spanish Empire.[44]

Falciparum malaria from African slaves similarly transformed the European colonies in the West Indies and Carolinas, where local mosquitoes and climate were amenable to year-round transmission. As the number of African slaves in the West Indies increased, so, too, did the number of deaths among the Europeans who lived there. In the West Indies during the 1650s, English planters died three times faster than their new babies could be baptized.[45] European immigration, in the face of the death toll on the islands, ground to a standstill, and those who could do so made their fortunes as quickly as possible and then escaped.

Surviving letters and diaries from colonists in the Carolinas, after African slaves started disembarking in force, describe the arrival of a dreaded, deadly malignant fever. In 1684, one boatload of would-be settlers—warned by their ship captain that only two of the thirty-two "vigorous" people he'd previously carried from Plymouth to Carolina had survived their first year there—turned back before setting a single foot on Carolina's plagued coast.[46]

In 1685, a band of Irish settlers arrived in Carolina hoping to gather timber to take to Barbados. They all sickened with fever. Twenty-nine died.[47] A party of 150 Scots who arrived after a ten-week journey similarly abandoned their attempts to settle the area. "We found the place so extrordinerie sicklie that sickness seased many of our number," one wrote, "and took away great many of our number and discouraged others, insomuch that they deserted us when we were to come to this place."[48] Two young men who fled Carolina in 1687 arrived in Boston "pitiable to behold . . . They say, they have never before seen so miserable a country, nor an atmosphere so unhealthy," wrote a French settler who met them. "Fevers prevail all the year, from which those who are attacked seldom recover, and if some escape, their complexion becomes tawny."[49] Before they left Carolina, the refugees said, they'd seen a ship from London arrive with 130 people on board. By the time they left, 115 of those new arrivals were dead, "all from malignant fevers which spread among them."[50] The deadliness of Carolina's fevers found its way into proverb. "They who want to die quickly go to Carolina," said the English. Added a German commentator: "Carolina is in the spring a paradise, in the summer a hell, and in the autumn a hospital."[51]

Characteristic of communities suffering *P. falciparum*'s appetites, infant mortality skyrocketed. Eighty-six percent of European American babies born in South Carolina died before they reached the age of twenty. In one parish, over a third of European American infants died before their fifth birthday, with most dying in their first year between August and November, when the malarial mosquitos were biting. One typical South Carolina couple, married in 1750, bore sixteen children, of whom just six survived to adulthood.[52] Those who could afford to do so fled to the coast and the highlands during the late summer and fall, when malaria broke out most virulently, establishing the still-popular South Carolina resort communities of Summerville, Pawley's Island, and Sullivan's Island, among others. Medical authorities warned them not to return until after the first killing frost.[53]

Ironically, *P. falciparum*'s heavy toll on the European populations of the West Indies and North America's southern colonies deepened the colonists' reliance on the labor of enslaved Africans. Battered, overworked, underfed, and housed in miserable, filthy conditions, most of the enslaved Africans nevertheless possessed Duffy-negative red blood cells that made them completely immune to the Europeans' vivax malaria, and 30 to 40 percent carried sickle-cell genes and other inherited antimalarial blood cell deformities, which effectively protected them from the worst ravages of falciparum malaria.[54] European physicians marveled at African slaves' apparent resilience to the malarial fevers, such as the three-day-cycling "tertian intermittent," which swept away their own people. "I have not met among them with a pure tertian intermittent in the whole of my practice," an amazed doctor in the West Indies wrote, "and those of forty years' experience mention it as a rare occurrence."[55] "The white man is seen shivering with ague, his countenance cadaverous and his temper splenetic," a Carolina physician noted. But "the black, is fat plump and glossy, in the full enjoyment of health and vigor."[56] (This was surely an exaggeration: falciparum malaria killed sufficient numbers of infants that the sickle-cell gene circulated among generations of African slaves in Carolina as lushly as it did in West Africa.[57])

Planters in the West Indies and the southern colonies were willing to pay top dollar for slaves from Africa. West Indies planters would pay three times more for an African slave than an indentured European worker, with slaves "seasoned" to the local malarias attracting even higher prices than those newly arrived.[58] Southern planters would spend twice as much on a slave from Africa as on a native slave.[59] Thus the institution of African slavery thrived.

One wouldn't guess that malaria has much to do with the history of Scotland, tucked away in the cool, misty northern highlands.

But even there, in the waning years of the seventeenth century, the bounty of the New World beckoned. Scotland, a poor struggling nation, was banned from trading with the colonial possessions of its more powerful neighbor, England.[60] But the Scottish entrepreneur William Paterson, founder of the Bank of England, dreamed of Panama.

Unlike the Spanish, who considered a canal, Paterson imagined a road through the isthmus. "The time and expense of navigation to China, Japan, and the Spice Islands, and the far greatest part of the East Indies will be lessened by more than half," he mused, "and the consumption of European commodities and manufactories will . . . double . . . Trade will increase trade, and money will beget money."[61] With a road built across Panama, everyone could get a piece of the trade in sugar, slaves, silver, and spices that had made the Spanish Empire and Britain's East India Company so fabulously wealthy.

Paterson knew precious little about Panama from his own experience. But he'd read journals and books, and studied the maps and drawings of pirates, missionaries, and shipmasters.[62] The journal of the young pirate Lionel Wafer, in particular, inspired him as to the possibilities. Wafer grew up in the Scottish highlands and went to sea at age sixteen as a surgeon to East Indies merchants and, later, Caribbean pirates. After suffering an injury, he was left behind to recover in Darién, the eastern part of the Panamanian isthmus, where steep jungle-covered hills collided with the palm-fringed, white-sanded Caribbean coast. For several months, he lived in the jungle with the Kuna, who nursed, fed, and indulged the bedraggled pirate, painting his body and allowing him to sleep in their hammocks under their plantain-leaved ranchos.

Wafer was mesmerized. He wrote in his journal of the emerald forests full of fat, tasty monkeys, the rivers lined with cinnamon, sugarcane, and prickly pear, the sea teeming with sweet-tasting turtles and crabs. Thick honey and wax, free for the taking from sting-free bees, hung from swollen hives in the trees. Tobacco, plantain, yams, cassavas, and pineapples abounded, as did fabulous trees so useful to humankind that in just six months, a force of three hundred Europeans, Wafer mused, could fell enough to pay for an entire expedition.[63]

Paterson was sold. Scotland could become a world power to rival the Spanish and the English, he thought, if only this veritable Eden, this "door of the seas" and "key to the universe," could be pried open.

Wafer had noted in his diary that Panama was rainy. "'Tis a very wet country," he'd written. But the Scots knew about rain: over sixty inches of it poured over the country's rugged highlands every year.[64]

Their urban dwellers knew about disease: anyone who lived in Edinburgh had already survived a gauntlet of the illnesses of filth. A viscous slime of watery cow dung covered the city's streets, and lice were so ubiquitous that fine English gentlemen boasted that they never slept in the city without wearing their gloves and stockings.[65] Scottish settlers had already embarked on several New World ventures, too. Although none had secured a steady flow of riches for their hilly kingdom, they'd already set up colonies in New England, Canada, and Carolina.

And so perhaps Paterson's dream to establish a Scottish colony on the isthmus, and to build a road across its land bridge, did not seem so fantastic. When the English authorities whom he first approached for support declined to become involved, fearful of upsetting the Spanish and their own East India Company, Paterson took his proposal to England's northern neighbor. Paterson's scheme inspired the country, and thousands anted up to fund the venture, "even the poor and landless, the thieves, whores and beggars," the historian John Prebble wrote, raising four hundred thousand pounds sterling, almost a quarter of all the available capital in the small, rural land.[66] "The whole kingdom," sniffed Lord Thomas Babington Macaulay, "seemed to have gone mad."[67]

Paterson ordered fine ships to be built to order, and filled them with artillery, nails, tacks, candlesticks, tobacco, and saws and machetes (to knock down the Panamanian trees Wafer had written about). While the Scottish countryside descended into famine, Paterson's colonists stocked up on salt beef and dried cod, rum, brandy, and claret—a year's worth of provisions for their venture. They carefully packed all their most desirable things to sell to the other colonies in the Caribbean: thousands of wigs; tartan, muslin, and calico; tobacco pipes; and pewter buttons. To win over the local natives, they packed thousands of Bibles, and combs inlaid with mother-of-pearl. The twelve hundred colonists who would make the journey—including four hundred hardy warrior-farmers from the highlands—outfitted themselves in tartan hose, stockings, and wigs. Paterson promised each fifty acres of good farmland at Darién and ushered them on board the flotilla of ships, a copy of Lionel Wafer's sunny journal tucked into their bags.[68]

Only after they got out to sea did the colonists realize that somehow their famished Scottish vendors and packers hadn't provided a year's worth of beef and cod as promised, but just half that quantity, and even that was already starting to spoil. Forty of the twelve hundred died during the month-long crossing of the Atlantic, relatively few for a seventeenth-century voyage.[69] (It wasn't uncommon for three times as many to die on such a trip.)[70]

Once they passed the Tropic of Cancer, the trade winds that had rushed them across the ocean waned. The air grew hot and still. They stopped at some island colonies, and several passengers visited taverns. Undoubtedly, infected mosquitoes bit them, or stowed away on board. Soon a daily and horrifying spectacle unfolded.

Yellow fever was a special scourge of sailors. *Aedes* mosquitoes carried the virus and often preyed on ships. The virus descends rapidly, characteristically provoking terrifying black vomit. The infected either die or survive with complete, lifelong immunity. The virus takes what it can, then disappears as quickly as it arrived.

Yellow fever hit the Scottish voyagers with deadly force. Perfectly well in the morning, the infected colonists would spew black vomit and be dead by evening. The dumbfounded survivors heard the splash of three corpses hitting the water every day.

In time, the deadly scourge loosened its hold, and soon the bedraggled flotilla passed a cliff-covered spit of highland jutting into the sea. Cliffs lined one side; the other was bordered by mangrove swamps. The wind at their backs, the ships silently sailed into the narrow bay, carefully navigating around the submerged rocks at its mouth. They'd made it to the Darién coast of Panama, and on this high rocky peninsula, they'd build their settlement, New Caledonia.

Hungry, exhausted, and unnerved by yellow fever, they disembarked. While Paterson buried his wife, the others set about clearing land and building huts. Their start in Darién couldn't be called auspicious, but in fact their prospects weren't bad. The New Caledonians had survived the shipboard typhus, dysentery, and typhoid, and yellow fever, too. Now they stood upon an untouched land as rich in fish, fowl, turtles, and monkeys as Wafer had written, on the threshold of a great step forward for their homeland. With just a few days' work, a band of settlers could capture dozens of turtles—enough to feed more than a thousand men.

And yet they languished. The turtles and monkeys went unscathed as the colonists lay miserably in makeshift, insect-infested huts by the swamp, pathetically relying on the scarce and rotten supplies they'd brought from Scotland. For days they recorded nothing in their journals except the fact of rain: "much thunder, lighting and rain . . . great showers of rain . . . a prodigious quantity of rain . . . much wind and rain . . . wind and rain as above." They drank prodigious quantities of alcohol, wrote in their journals of sad dreams of plundering for gold and fleeing Darién in shame, and buried another eleven of their fellow settlers. Months later, they'd broken no ground, and their fort and town lay half-built. When a small Spanish party encroached on the little colony, three quarters of the colonists were too sick to fight.[71]

This wasn't yellow fever. The deaths and sickness likely stemmed from a variety of infections, including falciparum malaria. Expeditions to the area that suffered similar losses in later times—when the presence of *Plasmodium* could be discerned by its vulnerability to antimalarial drugs or by microscopic scrutiny—suffered demonstrably heavy burdens of *P. falciparum* infection. The parasite had many possible venues of access to the Scottish colony. It could have hitched a ride in a Scot drinking at a tavern during earlier stops in the Caribbean colonies. Or perhaps it had arrived with the Kuna leaders who promptly and regularly visited New Caledonia. Several colonists accepted invitations for feasts and overnight visits to open-air Kuna villages that offered no barrier to bloodthirsty mosquitoes, and where hammocks swung just a quick buzz from those of escaped African slaves. Richard Long, sent by the English to recover sunken silver ships off Darién in the late 1690s, writes of encounters with "a Spanish Negro who was slave" to local Kuna leaders.[72]

Day and night, the colonists wrote in their anguished letters home, mosquitoes harassed them in their dank little thatch huts.[73] "The air is abominable," one wrote home, "and the Water Poyson."[74] Paterson, delirious and incapacitated with bouts of fever, wrote that his fellow colonists "daily grow more weakly and sickly." "I was troubled with fever and ague that I raved almost every day," one colonist wrote, "and it rendered me so weak that my legs were not able to support me . . . our bodies pined away . . . we were like so many skeletons." Between January and May, nearly four hundred of the settlers died, with as many as ten to twelve deaths a day.[75] "Halfe [*sic*] of the men in the Collony [*sic*] are al seeke at present with the great heats . . . Our fortifications are not done as yet neither wil be these twelvemoneths," an ensign wrote to his family.[76]

They needed more oatmeal and cheese and brandy, they pleaded in letters sent back to Scotland. "I must be a beggar at present," an ensign wrote to his father, "thogh [*sic*] I hope not to be long so."[77] But the English decreed that none of their subjects in the Caribbean should help the withering colony, with its abominable air and poison water, and the few ships that visited New Caledonia exhibited a justifiable lack of interest in the Scots' wools and laces.

In June, with news of another Spanish attack looming, the colonists fled New Caledonia. Those too weak to make it to the boats were derided as "poor, silly fellows" and left behind to die. One ship made it to Jamaica in seven weeks, losing 140 lives along the way. Another arrived in New England, losing 105.[78] Paterson lay in his cabin unable to move, his dream in pieces.

Later historians made much of the great profusion of brandy bottles that littered the colony's ruins, half-jokingly suggesting that the Scots had squandered their colony on a long drunken party, but the truth is that they'd had no better medicine—besides opium—to fight Panama's malaria.

• • •

By the time the sole surviving ship of the Scots expedition to Darién made it back home, Scotland had descended further into famine, and three hundred hungry new settlers, inspired by the land of plenty described in Lionel Wafer's recently published journals, had already set sail for Darién to join the others. Goldsmiths, distillers, ministers, and the wives of the colonists sailed off to Panama hoping to civilize New Caledonia. While they stalled around the Isle of Bute, they heard that the earlier colonists had abandoned the settlement, but dismissed the news as vicious English gossip. Who could have taken out those hardy souls, endowed with all the best assets of Scotland?

When the news became unavoidable—letters from the terrified Caledonians had started to arrive—the directors of the expedition remained stalwart. The earlier colonists had "shamefully deserted" and the latest expedition was to "repossess yourself [of New Caledonia] thereof by force of arms." The first settlers had conspired with their northern rivals, obviously, and betrayed the Scots. Nobody could imagine there might be a worse, and much tinier, enemy than the English.

When the second wave of colonists arrived in Darién, they found the colony in ruins, discovered the sad scrum of gravestones. They, too, started to go hungry, forced to cut rations when one of their ships caught fire and sank with all its provisions after someone on board tripped over a candle while getting a nip of brandy. They, too, started to fall ill. Soon, two thirds of the second wave of Scottish colonists were sick with fever.[79]

"I was brought to the gates of death," one of the chaplains on the colony wrote in a typically sad letter to Scotland,

> by a long and severe fever . . . and about 150 persons were cut off by death, beside what have dyed since . . . instead of the comfortable settlement we expected, nothing left but a vast howling wilderness

> in the properest sense . . . here all the circumstances of unpassable woods, vast desolations never frequented by mankind, Retired Recesses and resting places of tigers, Buffaloes, monkeys and other wild beasts, all manner of dangers and difficulties . . . we are a poor graceless shiftless and heartless company labouring under all discouragements: having no lodging, but either on board the ships . . . or under the shadow of trees in the woods or little huts made of the branches; and not provisions but what we brought from Scotland, which is now musty, rotten, old and salt, and yet like to be very soon exhausted; which if we get no supply speedily from Scotland will reduce us to the greatest extreamity.[80]

Slowly, Spanish soldiers—fighting off an outbreak of fever themselves—surrounded the weakened colony.[81] Like those before them, the new New Caledonian leaders didn't even bother to organize a defense. Hundreds were ill, and another accidental fire burned down more huts.[82] The Spanish demanded a surrender, and with more than a third of the New Caledonians too sick even to stand, and sixteen a day falling into their graves—"our Fort indeed like a hospital of sick and dying men" one later wrote—they did.

Crowded onto their leaky boats, even more New Caledonians succumbed to the "malignant fevers and fluxes . . . [that] swept away great numbers from among us," a survivor noted. "They would sometimes bury in the sea eight in one morning . . . and when men were taken with these diseases, they would sometimes die like men distracted, in a very sad and fearful-like manner."

Of the four thousand Scottish colonists who set out for Darién in the late seventeenth century, two thousand had died. Survivors trickled into New England and the West Indies. Just a handful made it back to Scotland. Their disappointed relatives and neighbors considered these hardy souls—who had survived typhoid and typhus, yellow fever, malaria, and the Spanish—personae non gratae. Angry mobs surrounded them and derided them as weak and cowardly. Their own fathers felt too ashamed to see them. "They were a sad reproach to the nation from which they were sent," wrote Francis Borland, a chaplain of the Church of Scotland.[83]

Thousands of Scots found themselves bankrupted by the Darién misadventure.[84] William Paterson ended his life in ignoble obscurity, teaching mathematics to poor children.[85] England offered to repay Scotland's debts, in exchange for its forfeiting its national autonomy to become part of a new Great Britain. The Scots accepted, and the independent nation of Scotland was no more.

The force of malaria's tide rumbles to this day. Descriptions, reenactments, and artifacts from the Scots' disastrous foray into Darién fill the libraries and museums of Britain. Malaria continues to repel foreigners from Africa. Standing on the coast of Cameroon you can see a gray blob shimmering on the horizon: Bioko Island, in the Gulf of Guinea. In the early 2000s, the Ohio-based Marathon Oil company built a giant natural gas liquefaction plant on Bioko, along with hundreds of neat ranch houses of the type that are commonly found in Texas. The ranch houses are supposed to be full of American oil workers and their families, but they are all empty. "Too much damn malaria," a malariologist who consulted for Marathon says.[86] Nobody wants to come and risk the bite of a local mosquito.

It's not easy to find written record of malaria's foray into North America. Some of the most comprehensive explorations have long fallen out of print. You cannot order Erwin Ackerknecht's 1945 analysis of malaria in the Upper Mississippi Valley, St. Julien Ravenel Childs's 1940 book on malaria in the early Carolinas, or Gordon Harrison's masterful 1978 study, *Mosquitoes, Malaria, and Man*, from any bookstore. You can't even find them at your local library. I acquired a beaten-up copy of Ackerknecht's book from the rare-book vendor Alibris. Seminal works by the malariologists who tackled American malaria, such as Paul Russell and Lewis Hackett, are even harder to find. The last remaining copies, it seems, gather dust in the noncirculating collections of university medical libraries.

I won Harrison's book on an eBay auction. It wasn't much of an auction. I think I was the only one who bid for it. It arrived wrapped in library-edition plastic, with a call number on its spine and the words DENNY JR. H. SCHOOL stamped on its cover. Apparently, the librarians at Denny Junior High, a public school in Seattle, had decided that even the nation's twelve-year-olds would not bother to check out this thoughtful history of malaria and had purged it from the collection. More damning evidence of authorial oblivion can hardly be imagined.

But grasping malaria's American legacy does not require a book so much as a map, for the disease is impressed upon the culture and demography of the United States. Disparate malarial burdens across the nation created deep cultural prejudices and settlement and demographic patterns that persist to this day.

Regional biases were born. Northerners, who suffered relatively little malaria, considered Southerners, with their endemic malaria, "voluptuary," "indolent," "unsteady," and "fiery," as Thomas Jefferson put it in a 1785 letter. (Northerners, by contrast, were thought to be cool, sober, laborious, and persevering.)[87] Racial biases also emerged. By the late eighteenth century, with *P. falciparum* restricting the growth of the European population, people of African descent had coalesced majorities and near-majorities in the southern colonies. Africans comprised 40 percent of the population of the Chesapeake colonies,[88] and the majority of the Carolina colonies.[89] As the slaves started to outnumber the European planters who owned them, the threat of slave rebellion loomed, and the novel and peculiar notion of solidarity based on skin color evolved, allying white planters of divergent class and ethnic backgrounds against the feared black majority.[90]

The bodies of black people were considered "tinctured with a shade of the pervading darkness," as a prominent Louisiana physician wrote in 1851,[91] and blacks were on "the lowest point in the scale of human beings," as the Alabama physician Josiah Clark Nott put it.[92] Malaria was considered tolerable for blacks, and intolerable for whites. "Negroes" were "lower animals," the American malariologist Lewis Hackett wrote in 1937, who could withstand malaria, while "human beings of the white race" could tolerate no malaria whatsoever.[93] In part, these notions stemmed from misunderstandings about Europeans' and Africans' different immune responses to falciparum malaria; they also rationalized the casual disregard that a racist culture propagated.

Today, attitudes of Northern superiority and white supremacy may have softened a bit, but the dense population and relative prosperity of the North compared to the South remain. Population density in some of the northeastern states today rivals that of India and Japan.[94] We live cheek by jowl, importing nearly all our food. In Alabama, by contrast, there are but thirty souls per square kilometer, a dispersal of humans more similar to, say, Madagascar than to New Jersey. As established during malaria's reign, today's African American populations are larger than average, and the economy relatively impoverished.[95]

Drive through northern Alabama, as I did a while back. One passes thick forests, burbling brooks: the very picture of fertility and easy living. And yet the roads are empty. The motel I checked into was unoccupied, the restaurant vacant, the expansive Wal-Mart parking lot only lightly used. For hours, the only evidence of human habitation is a hand-lettered sign that reads like a paranoia-tinged shriek from the solitude: "Go to Church, or the Devil will get you!"

4. MALARIAL ECOLOGIES

Malaria is not a disease of the environment in the way that, say, asthma is or certain kinds of cancer are. And yet its transmission depends upon an exacting set of environmental conditions. The protozoan parasite, despite all its sophisticated wiles and cunning, is more like a seed than a self-sufficient predator. Like a pip on the wind, it must alight in a fertile bed, be enveloped in the proper amount of moisture, and be bathed in the correct level of sunlight. The right mosquito must bite at the right time and with the correct frequency. If a local mosquito bites the wrong host, or if the insect's body becomes too cool or too warm, or if it dies or fails to bite before the parasite has time to develop inside its body, *Plasmodium*, one of the world's most deadly pathogens, might as well be an inert gas.

The circumstances that decide malaria's fate are contingent upon other circumstances equally beyond the parasite's control. The biting behavior of the mosquito, for example, depends partly on the species and partly on the variable availability of blood-filled hosts. Some species, such as *Anopheles gambiae*, are deeply connected to human hosts. Others are not so picky and will happily feed on the blood of cows or horses, if these happen to be available. The longevity of a mosquito, too, depends on multiple factors, such as the mosquito's habits and where she takes her blood meal. The female must rest soon after a feast, to excrete the excess liquid from her body. Will this siesta occur nearby, in a safe, snug place, or will it require some dodgy flight, in which she crosses paths with a swatting hand or swooping predators? What kind of weather will the blood-engorged insect encounter? Arid conditions, for example, can be deadly.

Of all the micro-geographic and climatic forces affecting malaria, the single most important factor is the species of the local population of mosquito. Of the planet's 430 different species of *Anopheles* mosquito, some 70 species transmit malaria. Each specializes in a specific geographic and climactic zone, be it the temperate Americas or the Asian tropics. For example, you won't often find a tropical African *Anopheles* in Northern Europe. But within each zone, you will find perhaps a handful of different *Anopheles* species, some of which are unreliable malaria carriers, while others are fabulous at it.

It would be nice if the variety of local species of *Anopheles* depended on some unchanging factor in the landscape, so we could simply avoid those places where the worst mosquitoes lived, just as we avoid living around, say, alligators or grizzly bears. Unfortunately, the peculiar mix of species in a given locale depends mostly upon a rather more mutable part of the landscape. The impregnated female mosquito must lay her eggs in bodies of water where they will hatch and feed on whatever debris floats by. The larvae's survival depends on being deposited in an amenable place to which its kind has been specifically adapted. Some thrive in salty water; others must have fresh. Some require shade; others, sun. Some demand flowing waters; others, stagnant.

The trouble is that the hydrology of puddles, streams, and ponds is one of the more mercurial aspects of the environment, vulnerable to any number of disruptive influences. We remake mosquitoes' microhabitats ourselves, mindlessly and routinely, by felling a few trees or digging a few holes. In so doing, we alter the temperature, rate of flow, and chemical composition of puddles, streams, and pond edges. To us these small alterations seem like nothing. But for the mosquitoes, they are the difference between life and death.

When the landscape is static, there's only so much of each kind of larval habitat available, and the mix of local *Anopheles* species can thus remain relatively stable. All things being the same, the larvae of the dominant species will fight off any intruders, and its populace is likely to become increasingly adapted to its specific niche. Once malaria transmission is established, the local people will, too, in time grow accustomed to the parasite, acquiring a patina of partial immunity. A relatively stable malarial ecology is established. Mortality declines.

But when the local malaria ecology twists and turns, new opportunities arise for the malaria parasite. Perhaps the local vector's habitat is extended, allowing the parasite to reach into new human populations. Or maybe the local vector is crowded out and a new, more efficient *Anopheles* population takes root, allowing the parasite to penetrate the local humans in more robust ways. Then the parasite's gains are the local humans' loss, for it can adapt much more quickly to the changed conditions than can the humans upon which it preys. In the lag between exposure to the new pattern of malaria transmission and the acquisition of immunity, the death toll rises.

Take the Roman Empire. The founding of the ancient city of Rome upon the banks of the Tiber in 753 BC created a prime habitat for the European malaria vector, *Anopheles atroparvus*. The Tiber, an actively migrating stream back then, regularly flooded its banks, leaving behind scores of puddles and pools in which *A. atroparvus*'s young thrived.[1] The Roman penchant for vegetable gardens, fountains, and impluvia provided even more mosquito nurseries. With an abundance of available larval sites and plenty of Romans to feast on, *A. atroparvus* abounded in Rome.[2] By 200 BC, a stable malarial ecology had been established, with *A. atroparvus* regularly passing on *P. vivax* parasites to the locals.[3]

Luckily for Rome, while *A. atroparvus* ably carried *P. vivax* parasites, that mosquito wasn't a reliable carrier of *P. falciparum*. Unlike *P. vivax*, *P. falciparum*'s survival depends on continuous transmission, without which it dies out, trapped inside its host. A stream of *P. falciparum* parasites regularly trickled into the Italian peninsula, in the bodies of traders and slaves from Africa. But *A. atroparvus* successfully foiled it before it took root. For one thing, the *A. atroparvus* mosquito is as attracted to animals for its blood meal as it is to humans, so its carriage of malaria to the correct host is not the most reliable. Every now and again, a falciparum-infected *A. atroparvus* mosquito would deposit *P. falciparum* parasites inside a cow or horse, which meant certain death for the parasite. Worse, *A. atroparvus* hibernates all winter. Once the cool weather arrives, it stops biting and repairs to some dark, warm corner for weeks at a time. For *P. vivax* parasites, which can go dormant, this wasn't a problem. But pauses are a deal-breaker for *P. falciparum*. After a few weeks inside the body of a human or an insect without access to new blood, *P. falciparum* parasites disintegrate.[4]

More reliable, nonhibernating, human-loving biters such as *Anopheles labranchiae* flit in North Africa, on the other side of the Mediterranean. Ancient Rome increasingly relied on imported grain from the second century onward. Stowaway *A. labranchiae* would have regularly arrived in Rome on the grain ships from North Africa, and were able travelers. While traders loaded the ships, rain showers might fill some broken clay jars with water, into which a passing female *A. labranchiae* might lay her eggs. The tiny ornamented pods, coated in a velvety pile, would balance upon the water's surface with two air-filled floats shaped like fans on either side.[5]

If conditions were amenable, by the time the wormlike larvae, with their giant eyes and spiky whiskers, hatched, they would have been in Rome.[6] Nobody would have noticed the skittish, ascetic little pupae they became. They don't eat anything; they have no mouths. If even so much as a shadow passes over them, they flip their tails and duck out of sight, rising to the surface again only to breathe. When the adult *A. labranchiae* emerges from the pupae, she is soft and wobbly. But after half an hour, her cuticle stiffens and she flies off to some still, dark corner. She can fly as far as eight miles in search of a meal. With the help of a strong wind, she could end up hundreds of miles away.[7]

But Rome's *A. atroparvus* effectively repelled such interlopers, and had already claimed all the best mosquito nurseries, those watery areas free of predatory fish. If any *A. labranchiae* mosquitoes successfully deposited a few eggs somewhere, they'd be fish food soon enough. If they dared lay claim to *A. atroparvus* turf, there'd be dire consequences. *Anopheles* actively guard their territory from encroachment by rival species, their larvae secreting deadly chemicals to kill off any newcomers that they don't devour.[8]

The ecology of malaria in early Rome was thus both stable and resilient. This helped strengthen the empire, for it meant that the Romans had ample opportunity to adapt to life with the parasite, and exercise an immunological advantage over foreign intruders who did not. Most powerfully, people across the Italian peninsula and around the Mediterranean developed genetic defenses against the worst ravages of the parasite. Genes that disrupted an enzyme called G6PD, required for normal functioning of red blood cells, emerged and spread. The defect impaired human bodies' ability to repair oxygen damage, so that malaria-infected cells essentially poisoned themselves. (The main drawback: the peninsula's famous fava beans could send G6PD-deficient Italians into a spiral of hemolytic anemia, a condition known as favism, after the bean.)[9]

The Romans adapted culturally, too. They understood enough about the epidemiology of their malaria to minimize exposure to it. Ancient Roman scholars such as first-century BC writer Marcus Terentius Varro warned that animals too small to be seen (he called them *bestiolae*) entered the mouth and nostrils and caused horrible diseases.[10] He recommended that Roman houses be built on high land, where the wind would blow the beasties away. Roman elites thus built their sumptuous villas in the mosquito-free hills. Even the peasants who worked the infested lands below the villas knew to avoid the sickly winds, building their houses with windows facing in, toward a central courtyard, rather than out into the breeze.[11] The very worst mosquito-ridden regions, such as the rich wetlands of the Pontine marshes and the Roman Campagna, which ringed the metropolis, were abandoned to brigands, highwaymen, and the odd pallid peasant. Although these were the nearest and best agricultural lands, to avoid their malarious mosquitoes, the Romans (after a period of development) left them sparsely settled.[12]

For the inevitable infections they suffered, the Romans devised a varied and fanciful welter of antimalarial therapies. The malarious might try some honeysuckle dissolved in wine to relieve their swollen spleens, or perhaps consume the liver of a seven-year-old mouse.[13] They might, as the emperor Caracalla's physician, Serenus Sammonicus, recommended, wear a piece of papyrus inscribed with a powerful incantation—"abracadabra"—around their necks as an amulet. Bolder souls might try Sammonicus's other malaria cure: bedbugs eaten with eggs and wine.[14] They might try waking at dawn three mornings in a row, facing a window, and shutting it suddenly while reciting a prayer, or, for a male sufferer, having intercourse with a woman just starting to menstruate.[15] Prominent Roman physician Galen and medical scholar Celsus advocated energetic bloodletting. Finally, when all else failed, the Romans prayed for relief to the demon goddess of malaria, Febris, in three dedicated temples around the city.[16]

Just as malaria immunity helped the Bantu spread and restrained European intrusion into malarious Africa, so Rome's cultural and biological adaptation to chronic vivax malaria helped strengthen the city's ability to repel outsiders. To enter the capital, the armies of malaria-free Northern Europe would have to spend days in the malarious swamps around the city, exposed to the bites of infected *Anopheles*. Unlike the regularly exposed Romans, the Northern Europeans had no tricks to help them minimize malarial feasts on their bodies. Time and again foreign armies fell prey to Rome's malaria. "When unable to defend herself by the sword," the poet Godfrey of Viterbo noted, "Rome could defend herself by means of the fever."[17]

And so, even as Julius Caesar lay in bed with malaria, his armies conquered lands far and wide, bringing booty and slaves to enrich Rome.[18] Having forsaken its best farmland to malaria, ancient Rome could not feed itself, but the riches of conquest paid for grain, olives, fish sauce, and oil imported from North Africa. It paid for elaborate aqueducts, allowing wealthy Romans to move away from the most mosquito-ridden water's edge.[19] Until the environmental conditions that underlay Rome's stable malarial ecology shifted, the malarious Roman Empire thrived.

But building an empire required natural resources, and Rome's oak forests suffered the brunt. As the peninsula was deforested, the usual changes occurred. Erosion intensified. As sheets of rainwater washed off the shorn hills and into the valleys below, the water table rose. Rivers flooded more easily. The plains grew marshy.[20]

What this meant is that Rome increasingly harbored a new, uncolonized mosquito habitat, and at some point, stowaway *A. labranchiae* from North Africa must have found amenable spots unharassed by *A. atroparvus* and quietly laid their eggs.[21] Adapting to Rome's cool winters required a simple adjustment for the nonhibernating insects—spending the winter indoors, say, inside warm, dimly lit Roman homes, where they could continue to bite year-round.[22] It wasn't a big stretch. The behavior of *Anopheles* mosquitoes is not that rigid. We don't know precisely when *A. labranchiae* made itself at home in Rome, but we do know that it did. In time, the mosquito established a foothold on the peninsula as well as on the islands of Sicily and Sardinia, which *A. atroparvus* had failed to colonize.[23]

By the fifth century AD, Roman villagers were suffering horribly from *P. falciparum* outbreaks. The disease terrified them in a way that suggests the scourge had been previously unknown. Between 1988 and 1992, the archaeologist David Soren and his team from the University of Arizona excavated the remains of nearly fifty infant corpses from a fifth-century village near Rome called Lugnano. The infants had died in rapid succession and been buried hastily, in an ad hoc trash heap, entombed with mysterious offerings to pagan gods. Along with the bodies of their dead babies, the villagers had buried the torn-off jaw of a six-month-old puppy and the carcass of a dog split straight down the middle. They'd included the claws of ravens and singed honeysuckle branches. The hands and feet of the body of a two-year-old child had been weighed down with giant stones and tiles.[24]

The carnage started to make sense when molecular biologists discovered the DNA of falciparum parasites inside the excavated bones.[25] *P. falciparum* preys most prolifically on babies and young children, which would explain the predominance of dead babies at the site. After the first few infections, falciparum malaria spreads quickly—a single victim can infect one hundred others. Thus the quick succession of hasty burials. Honeysuckle was a known Roman salve for malarial symptoms. As for the mutilated dogs—well, if it was *P. falciparum* that killed the babies, the outbreak would probably have occurred in the heat of summer, when the waters of the Tiber running alongside the village receded, leaving behind a mess of puddle and marsh. Those long summer days were known in Roman times as the *caniculares dies*, the dog days, when the dog star Sirius disappears in the glow of the sun. And the keeper of infants' souls, according to Roman mythology, was the goddess Hecate, who sailed through the heavens on a chariot pulled by the hounds of hell.[26]

Making offerings to appease an enraged Hecate made sense in the face of a summer outbreak of an infant-slaughtering pathogen. Early medieval Romans had no other explanation for an illness that suddenly struck so many people at once. Their medical authorities considered sickness to be the result of idiosyncratic imbalances in the body. They had no concept for contagion. A curse cast by the pagan gods of their ancestors must have seemed the most reasonable explanation, the goddess Hecate the most likely culprit.[27]

But the canine sacrifices also point to the terror the villagers must have felt. Practicing pagan rituals in that period risked serious political consequences. By the fifth century, Christianity had been the official religion of the Roman Empire for some two hundred years.[28] The Church reserved special venom for the followers of Febris, and anyone caught wearing an amulet was to be executed.[29]

It wouldn't have been the fact of child deaths that scared the Romans. Generally speaking, fewer than half of the infants in early medieval Roman villages such as Lugnano survived childhood, and even the adults wouldn't have expected to live past twenty years.[30] More likely, the deadly epidemic was something novel, something they'd never seen before. As falciparum infection progresses, the symptoms of fever and chills become much more pronounced, and divergent from the usual vivax malaria to which the villagers were undoubtedly inured. Some victims would have fallen into open-eyed comas and gone into convulsions. Such symptoms might well have struck the villagers as otherworldly, which would explain why they weighed down the body of one dead child with stones and tiles, as if to prevent the evil spirit that seemed to possess her from rising again.

If the fifth-century falciparum outbreak in Lugnano was indeed new, its emergence in Rome coincided with a broader dissipation of the empire's power. The fifth century found the empire weakened, under attack, and wasted by famine. In AD 401, for the first time in eight centuries, the city of Rome's famous defenses were breached by Alaric and his Visigoth army.[31] Fifth-century Romans had to survive on shipments of food from North Africa, a thin thread that northern armies severed simply by holding up the grain ships at sea, plunging Rome into famine.[32] The elaborate villas that sat above villages such as Lugnano lay in ruins; villagers squatted in the rubble.[33]

Historians still debate what triggered Rome's decline. Was it the empire's internal contradictions, the superior technology of its rivals, its trade deficits, its plagues and pestilences? Clearly, many things went awry. But the transformation of malaria from the fever that protected Rome to one that killed, occurring just around the time of Rome's decline, surely exerted a demoralizing and destabilizing effect.

As falciparum transmission became established, the troublesome but mild malaria season would have yielded to a year-round scourge, which struck foreigners and Romans with equal severity. *P. falciparum* would have laid bare the inadequacies of Roman medicine and mythology. Amulets and prayers might have seemed effective in the face of self-limiting vivax infections, for by probability alone, their use would have sometimes coincided with *P. vivax*'s natural cessation. Not so with *P. falciparum* infection. Its arrival, thanks to slow-moving ecological disruptions, shattered all the old certainties.

By AD 476, the Roman Empire was no more, its canals filled with rubble and its aqueducts crumbled.[34]

Foreign powers took control of Naples, Sicily, and Sardinia. Only the northern city-states, beyond *A. labranchiae*'s reach, prospered. Elsewhere, *P. falciparum* took so many lives that the deadly compromise forged in Africa—the sickle-cell gene—emerged and spread along the shores of the Mediterranean.[35] Even the most celebrated Romans, such as the poet Dante Alighieri, suffered the "shivering of the quartan."[36] Dante died of malaria in 1321.[37]

The Vatican founded a vast hospital, Santo Spirito, along the banks of the Tiber, which overflowed, generally to at least three times capacity, with fever patients.[38] But there wasn't much that could be done to save the sufferers. The parasite took the life of Pope Innocent VIII in 1492, Pope Alexander VI in 1503, Pope Adrian VI in 1523, and Pope Sixtus V in 1590.[39]

The people of Rome no longer could understand their fevers. They seemed to have something to do with bad air, they said: the *mal' aria*.[40] The demons of air, water, and earth were locked in battle with the demon of cold, the sixth-century historian John Lydus speculated.[41] No, a foul dragon lurked in a cave beneath the city, others said, breathing out the bad air.[42] It was a vengeful Febris, the poet Poliziano said, flying through the air in her lion-drawn chariot, followed by a train of monsters. She injected flames from her torch and icy snow mixed with venom into the bones of her victims.[43]

Rome's unknowable malaria inspired images of horror still potent today. There's nothing intrinsically sepulchral about mists and wetlands. And yet, then and now, writers describe these environments as deathly. "The rising of the sparkling Dog Star at the morbid foot of Orion was imminent," one medieval bishop wrote, in anticipation of a deadly malarial summer in Rome.

> All the air in the vicinity became dense with misty vapours arising from the neighbouring swamps and caverns and the ruined places around the city, air that was pestilential and lethal for mortals to breathe . . . the rage of the Dog Star . . . grew even hotter, and there were hardly any men left who were not debilitated by the seething heat and bad air.[44]

Healthful northerners visited the malarious Vatican and the ruins of Rome and professed disgust. "There is a horrid thing called the malaria, that comes to Rome every summer, and kills one," Horace Walpole wrote in a 1740 letter, introducing, at long last, the word *malaria* into the English language.[45] There was "a strange horror lying over the whole city," wrote the English critic John Ruskin in 1840. "It is a shadow of death, possessing and penetrating all things . . . you feel like an artist in a fever, haunted by every dream of beauty . . . but all mixed with the fever fear."[46]

"The Valley of the Shadow of Death"—that's how Florence Nightingale, in 1847, described the silent, thyme-covered Roman Campagna into which nonimmune villagers from the surrounding hills descended during the summer to harvest wheat.[47] The wheat ripened at the height of the malaria season, and the impoverished peasants who worked the fields spent their nights in caves, stables, and under the stars, easy prey for mosquitoes.[48] They were "the most unhappy, most resigned" people in Italy, the French writer Stendhal wrote in 1829. "They visited Rome on Sundays, dressed in their primitive costumes, their faces showing traces of malaria."[49] They were "pale, yellow, sickly," wrote Hans Christian Andersen in 1845.[50] In the first half of the twentieth century, it wasn't unusual for the women who stayed behind in the mountain villages to lose three or more husbands to the Campagna's fever.[51] As a result, two million hectares of arable land remained fallow, and two million more were "cultivated badly," writes historian Frank Snowden.[52]

The earth's axis wobbles about one degree every seventy-one years, so the dog star and the sun no longer rise as one in the summer sky. The dog days are technically over. But the plague of *P. falciparum* that befell Rome during the end of the empire and those early medieval *caniculares dies* lingered for more than a thousand years.

Of course, the kinds of environmental disruptions that allow more malignant malarial mosquitoes to extend their territory vary from locale to locale. In the northeastern United States, the troubles began when people started building dams.

Colonial New England's myriad brooks and creeks provided plentiful habitat for populations of *Anopheles punctipennis*, a little forest mosquito that thrives in shaded, running waters. *A. punctipennis* has a predilection for animals, and so it wasn't a particularly potent vector for the few malaria parasites it picked up here and there. Despite hot summers and the repeated introduction of vivax parasites from farther south, malaria's grip on the Northeast was weak and sporadic.[53]

But by the late eighteenth century, industrious New Englanders started to realize they could harness the power of the region's rocky, tumbling rivers to card wool, grind grain, and cut logs. All they had to do was build some dams so they could draw down the water power as their mills required.[54]

Behind the dams, of course, what was once tumbling river got backed up into still, sunny ponds. *A. punctipennis* faltered in such environments, but the sun-loving *Anopheles quadrimaculatus* thrived in them. *A. quadrimaculatus*, which unlike *A. punctipennis* happily entered homes in search of blood, was malaria's prime vector in the southern states. Now they started moving north. As New England's brooks turned into vegetation-choked ponds, *A. quadrimaculatus* populations grew and *A. punctipennis* populations declined.[55]

This ecological shift overlapped with the advent of the Revolutionary War, during which nearly half of some regiments were infected with malaria parasites. Infected soldiers returning home to New England introduced the parasites into a subtly but powerfully transformed landscape.[56]

One such soldier was Elijah Boardman, from New Milford, Connecticut. Boardman suffered weeks of fever during the war. For forty days, he sweated and shook while camped in the mosquito-ridden swamps of Long Island. The experience "left not less enduring traces," one of his descendants would later write, "then such wounds as he would cheerfully have received . . . from the musket-ball or sword."[57]

His father advised him in letters: "If you are not well," he wrote, "do not think too much about home."[58] It was good advice but, for Boardman and countless other malarious soldiers, impossible to follow. Rather than evacuate to New Jersey, as their superiors advised, they fled on wagon and horseback to their homes, malaria parasites burning in their veins.[59]

The town of New Milford, Connecticut, situated along the banks of the Housatonic River, centered around a muddy, manure-littered town green. Although he suffered regular bouts of malaria—most likely relapses of vivax malaria from his original infection during the war—Boardman prospered there. He opened a general store with his brother, selling lace, wine, and imported tea, and bought several farms and fishing rights.[60] The painter Ralph Earl arrived in 1789 to paint his portrait, which eventually made its way into the Metropolitan Museum (where it still hangs). His name grew to become a "passport to particular respect" in New Milford, as a local historian wrote,[61] and when a widely admired local beauty consented to marry him, Boardman began building one of the grandest homes New Milford had ever known, a great Georgian mansion that stands to this day.[62] The Housatonic River flowed gently behind the house, and was dammed not far off, powering two small mills that ground grain and cut logs.[63]

In the summer of 1796, that dam was raised ten inches by its new owner, Joseph Ruggles. The waters of the Housatonic above the dam flooded over its low-lying banks, swamping more than fifty acres of low ground behind Boardman's house. The water settled into a large, shallow pond.[64]

Even inside his Georgian mansion, parasite-carrying locals such as Boardman were vulnerable to the bites of bloodthirsty *A. quadrimaculatus*. The insect came silently and softly in the middle of the night, barely noticed. When she did, Boardman's war-era parasites, locked inside his veins by the capricious biting behavior of *A. punctipennis*, emerged out of dormancy, and took flight into the New Milford night.

Within weeks, malaria spread throughout the town. Three hundred New Milford residents fell sick with fever.[65] "Almost every family near the middle of the town have been afflicted more or less," Boardman informed his brother-in-law in a letter. Three of Boardman's employees had been taken down by the fever, and two townspeople had died. "So many persons are sick that it is almost impossible to get sufficient assistance from those that are well to take care of those that are ill," he wrote.[66]

During the last week of August, things deteriorated in the Boardman mansion. On Saturday, Boardman's wife spiked a fever. She was "considerably reduced," he wrote. "How long it will last or how low she will be brought cannot be known at present." On Monday, Boardman's two-year-old son, William, was down, too. On Wednesday, as his wife lay prostrate in her bed, Boardman watched William convulse. He scrawled a letter to his brother-in-law. The child was "very sick and decaying," he wrote in a shaky hand. Boardman's only hope was that some person, perhaps a doctor, perhaps his father-in-law—his usually fine handwriting, at this point in the letter, is now illegible—would arrive in New Milford to rescue them.[67]

Malaria broke out all over southern New England in those final years of the eighteenth century, everywhere in connection with the establishment of a local milldam. In 1795, it struck Sheffield, Massachusetts, about forty miles north of New Milford, where a dam had created a swampy pond. "A number of inhabitants, about the north pond [are] afflicted with a fever," remembered the local physician, Dr. Buel. "The people first attacked were those who lived nearest to the pond; whole families of whom were taken down at once." By the fall, two thirds of the population within three quarters of a mile of the pond were sick with a raging malaria. "The pains in the head, limbs and back were very severe," Buel remembered, and their faces and eyes had turned a terrifying yellow, suggesting the possibility that *P. falciparum* may have been at work.[68] A dam at South Hadley, which flooded ten miles of meadowlands, similarly rendered Northampton, Massachusetts, long considered "one of the healthiest" towns in the area, "extremely afflicted with fever and ague."[69]

Two years later, another malaria epidemic struck New Milford, and a year after that, another.[70] During one year's epidemic, nearly one hundred perished. "Young as you are, you may die," Boardman's wife warned their son in a letter. "Endeavor to be ready."[71]

A similar confluence of factors occurred in the wake of the Civil War, bringing another wave of malaria to the northeastern United States. The war itself served as a giant malarial feast. Union troops suffered 1.3 million cases of malaria, leading to 10,000 deaths.[72] In 1864, every single federal soldier in the Union army active in Louisiana and Alabama came down with at least one episode of malaria.[73] Over half of Northern troops suffered the scourge.[74]

War-making—the digging of trenches, the destruction of dams, the building of roads—levels an ecological insult that malaria can often exploit. Trenches fill with water, craters become puddles, previously untrammeled valleys become rutted and fetid. At the same time, war brings together great masses of previously unacquainted people, with their varieties of malaria parasites and immunities, in the middle of prime mosquito habitat. In countless wars, malaria has killed more soldiers than combat. And the intensity of wartime malaria can extend well into peacetime, as soldiers returning home introduce their newly gained malaria parasites into virgin landscapes, triggering yet more malaria epidemics.

As the Civil War soldiers returned home, outbursts of malaria rippled northward from New Jersey to New England. Madison Square, Washington Square, and Tompkins Square in Manhattan became "dangerous hot-beds of disease and death," as a *New York Times* headline put it.[75] Every man, woman, and child in the neighborhoods of Dutch Kills and Ravenswood in Long Island, it seemed to a *New York Times* reporter in 1877, had been "poisoned" with malaria. "There has been so much malarial fever that it amounts almost to an epidemic," the *Times* reported. The schools were emptied of students, and half the police force was "unfit for duty." Residents fled the island en masse, "To Let" signs fluttering on their abandoned homes.[76]

In Bound Brook, New Jersey, not a single family escaped malarial infection. "I have resided here 33 years," a lumber merchant told a newspaper reporter, "and was never compelled to take a dose of [the antimalarial remedy] quinine, or use it in my family, until 1878. Now we all take it in pretty large quantities, and have had touches of the malaria in some form."

Across New England, the story was the same: chills and fevers reported in epidemic form all the way up to the foot of the Berkshire Hills.[77]

• • •

Malaria's victims sensed that their plight had something to do with the changing landscape, in particular the recent spread of still waters. Over the previous years, as New York City had grown to become the nation's largest metropolis, Manhattan Island's watery idyll of stream, creek, and bog had been paved over entirely. Critics suspected that the graded streets, by blocking the island's natural drainage, were to blame for outbreaks of fever.[78] Worse, city officials had constructed the city's public squares atop the swampiest parts of the island, land that private builders had rejected. "Unhealthy vapors" rose from the "stagnant and mephitic" waters below foot, critics charged.[79]

"Every case of death which occurs from malarial disease in an organized community is a crime committed by the authorities," one irate reader wrote to *The New York Times*. "I consider the neglect of the absolutely necessary precautions to preserve health by drainage as much a crime against humanity as the burning of Russian hospitals by the Turks."[80] Leaving New York City's underground streams intact was a "suicidal policy," the paper editorialized.[81]

"There is but one radical remedy for this scourge anywhere," proclaimed New York public health official General Egbert L. Viele. That, he said, was the destruction of milldams. "Mill-dams produce . . . vegetable decomposition over a wide extent of territory," he explained. "This decomposing vegetable matter must be removed and these mill-dams converted into what they ought to have been from the beginning . . . containing nothing but pure potable waters."[82]

But in the face of entrenched economic interests such arguments fell upon deaf ears. The nation's rising manufacturing sector relied on the pent-up water power the dams provided. In 1800, Connecticut manufacturers operated some fifteen hundred milldams in the state,[83] and by 1869, water power provided nearly 50 percent of all power used in U.S. manufacturing.[84] To protect the milldams and their operators, legislators passed "mill acts," which dramatically curtailed the damages locals could seek against mill owners whose dams had caused them flood or fever.[85] In 1805 a Massachusetts court went so far as to state that property rights themselves were null and void in the face of "all things necessary to the upholding of mills," which they described as "obvious and important purposes of public utility."[86]

In 1799, Elijah Boardman was reduced to leading a group of townspeople on a rampage against the dam in New Milford. The group headed into the fetid waters of Ruggles's millpond and, with their tools and bare hands, began to dismantle the dam piece by piece, until all that remained was some rubble on the banks.[87] The released waters rushed toward them, and millions of tiny black *A. quadrimaculatus* eggs washed away downstream.[88]

Critics of the milldams had better luck in the years after the Civil War, as water power faded into obscurity. Miners discovered the nation's rich veins of coal in the 1830s, and coal-fired factories rapidly eclipsed those powered by dammed water. By 1909, water accounted for less than 10 percent of U.S. manufacturing power; by 1919, just 6 percent.[89]

The mill owners' economic power sapped, public health experts stepped up their attacks on milldams. "I've had the supreme satisfaction of seeing a number of mill-dams destroyed through my own agency as an expert," Viele boasted. "One mill-dam of my own knowledge was the absolute cause of the deaths of over 20 adults." After five decades of man-made malaria, many milldams were destroyed and mill owners indicted for the public nuisance their dams had caused.[90]

The malaria outbreaks of the American Revolution and the Civil War that bled into the northeastern United States were far from the worst bouts of wartime malaria. The most notorious malaria epidemic associated with ecological disruptions and wartime population movements occurred during the First World War, on the Macedonian front.

The scene was the valley of the Struma River, which runs south from Bulgaria into Greece. Clusters of tiny sagging houses, verandas festooned with laundry, dot the hillsides, streams slipping down their sides. *Anopheles superpictus* roosted in these streams, while *Anopheles maculipennis* rose from the sodden, sun-dappled valley floor below. *P. vivax*, *P. malariae*, and *P. falciparum* parasites made a decent living, despite the local insects' fickle feeding habits. The villagers generously allowed their livestock to live in the lower levels of their homes, greatly improving the odds that the gourmand mosquitoes would eventually deposit parasites in the correct host. Malaria flourished in the valley six months of the year.[91]

In 1915, under the command of the French general Maurice Sarrail, six hundred thousand British, French, and Italian troops descended upon the Struma Valley. Their aim was to help the Serbs fend off the Bulgarians, but by the time they arrived, the Bulgarians had already beaten the Serbs. And so, in preparation for greater battles to come, they set up their tents along the spongy ground and started to remake the valley's landscape.[92] Needing roads for their motorized vehicles, they loosed thousands of locals upon the valley, shovels in hand, to clear brush and dig quarries, which the region's heavy rains promptly turned into a "giant's staircase of mud slides."[93]

Having thus extended the local mosquitoes' already capacious territory, the soldiers offered up their flesh, with the tents they retired to each night providing little barrier to the insects' entry. One soldier counted one hundred mosquitoes in his tent alone.[94] The introduction of just a handful of malaria parasites could easily have launched an epidemic, but, in fact, the bodies of the troops and the locals together housed scores of parasites from all over the globe. Each tent sheltered three soldiers, many of whom had arrived direct from other malarious fronts of the war. Local mosquitoes could pick up malaria parasites from India, East Africa, or Palestine, not to mention the already extensive range of local strains.[95]

"Malaria struck our men down like a scythe cutting grass," remembered one survivor. "In every battalion men went down by the hundred, and there were several cases of one or two officers and two or three score men . . . left out of a whole battalion up to full strength." The victims had to be dragged through the valley's still-trackless mud to the Greek city of Thessaloniki—which they called Salonika—on makeshift carts pulled by mules. Every afternoon, convoys of dozens of ambulances rumbled through the streets of Salonika to the general hospitals. "As they rolled silently along the busy, hot streets, one saw from behind each ambulance the feet of the four recumbent men within," wrote the British journalist H. Collinson Owen in 1919. Malaria sent nearly thirty thousand soldiers into Salonika's hospitals that summer, a flood of patients that outstripped the number of available beds by nearly three to one.[96]

Unaware of the epidemic, Allied leaders ordered General Sarrail to mobilize his forces for battle. Sarrail replied by telegram: "Regret that my army is in hospital with malaria."[97]

While the Bulgarians cut the railway line through Demir Hisar in Macedonia and captured the Greek port of Kavala,[98] the Allied soldiers fevered uselessly. Those who stayed out of hospital were not much better. The Salonika Army "was full of listless, anaemic, unhappy, sallow men whose lives were a physical burden to them and a material burden to the Army," wrote Owen. By 1917, Salonika's hospitals hosted more than sixty-five thousand troops sick with malaria,[99] the armies of three of Europe's most powerful nations "virtually paralysed," the malariologist Lewis Hackett later remarked, "before they could strike a blow."[100] And the sick soldiers were stuck there. With German submarines threatening to bomb hospital ships, a planned evacuation to Malta was scuttled.[101]

The mosquitoes continued to bite. After the extent of the malaria problem became apparent, the military organized anti-mosquito patrols to oil puddles and clear vegetation, but the Struma Valley's streams and marshes were under constant enemy surveillance and fire, and there was plenty of mosquito habitat beyond the patrols' reach, just over the front line.[102] And so when Greek hospital administrators rejected military health officers' suggestion that they install screens on the hospital windows, infected soldiers in hospital continued to reinfect others, and themselves.[103]

In the closing months of the war, an overland evacuation route safe from the German U-boats finally opened up. The officers at Salonika sent the sickest, most heavily infected soldiers back home, effectively relocating the Macedonian epidemic to the rest of Europe.[104] Five thousand fell ill as far north as the German coast, and in Archangel, Russia, in the Arctic Circle.[105] The old malarious counties of Kent and Essex in England suffered around five hundred cases.[106]

But malaria's First World War rampage did not stick in anyone's mind for long, overshadowed by the flu pandemic that struck in 1918. Only the neat rows of white crosses dotting the cemeteries of Salonika, marking the graves of Allied soldiers, bear testimony to malaria's World War I toll. The Salonika Campaign Society, dedicated to remembrance of the Salonika soldiers, still visits the graves annually.

Modern ships no longer ferry cholera vibrio from port to port. Or yellow fever. Physicians don't regularly infect patients with deadly bacteria. In most societies with sufficient resources, food preparation no longer spreads the pathogens of waste products. Public buildings do not broadcast tubercular bacteria.

But our mining, logging, and farming projects continue to disrupt environmental conditions in ways that create and spread malaria to this day. As late as the early 1990s, the World Health Organization complained that "economic development in agriculture and mining" continued to be a prime vector for the spread of malaria.[107]

Part of the problem is that some of the most desirable natural resources rest under prime malaria stomping grounds. Take the copper deposits buried under the Luanshya River, nestled between the Congo and Zambezi rivers in Central Africa. For years, fear of malaria kept both locals and outsiders away. Locals called the area "The Snake," for the seasonal wetlands that covered the area. Tall grasses, sedges, and rushes obscured their shallow, winding waterways, and the pathogen-carrying insects they harbored.[108]

But as the industrial revolution boomed, extracting the copper became increasingly alluring, despite the formidable microbes. The mining magnate Alfred Chester Beatty resolved to extract ten thousand tons of copper ore from Luanshya in the 1920s. The colonial government of Northern Rhodesia started building huts all along the Snake, most within half a mile of the feared wetlands.[109] Despite the offers of free room and board, Beatty's mining company had trouble recruiting sufficient workers. Some fled as soon as they arrived at the mine. Others worked for a week and then disappeared, not bothering to pick up their paychecks. When one worker, Joseph Zgambo, fell into the roiling river while assisting a surveyor, his fellow workers refused to work any longer. "They sat in a group muttering that the Snake had taken their fellow workman," the company's recruiter C. F. Spearpoint recalled, and the next morning they were gone. "The people are afraid that if they remain here they will certainly die," Spearpoint wrote. Within a few months, of the eleven hundred recruited workers, four hundred had fled.[110]

Those who stayed suffered the consequences. The construction of the mine and the township created new larval habitats for local mosquitoes daily, the malariologist Sir Malcolm Watson wrote,[111] and the mining company actively discouraged the use of local healers, who might have had some experience with the diseases common to the area.[112] People were advised not to waste their money on round-trip tickets to the mines. To add insult to injury, of the 500 cattle imported to help with transport, 498 died of sleeping sickness. Nearly every last dog died, too.[113]

More recent examples are not hard to find. Between 1970 and 1996, the Brazilian government, supported by the World Bank, engineered widescale development projects in the untouched jungles of the Amazon. Their agriculture and mineral extraction projects disrupted the jungle environment, creating new habitats for malarial mosquitoes. Migrant workers and others flooded into the region, residing in crude dwellings, where they were vulnerable to mosquito bites.[114] Soon, parasites from a sparse population of rubber tappers (unrecognized by the government), who traditionally lived in the jungle, started to infect the newcomers.[115] Between 1970 and 1999, the malaria caseload in the Amazon region of Brazil zoomed from around 30,000 to 600,000.[116]

Between 1983 and 1995, road builders, farmers, and others denuded more than four thousand hectares of Peruvian rain forest. Their new roads and fish farms extended the habitat of local *Anopheles* and brought them into close proximity to new, malaria-naïve settlers. More than 120,000 fell prey to *P. falciparum* in Peru in the late 1990s, compared to under 150 cases a year earlier in the decade.[117]

In the mid-1990s, encouraged by local government and international nongovernmental organizations, Ethiopian farmers replaced traditional crops with higher-yielding hybrid maize. Planting the maize required deep furrows in the ground, in which water collected and *Anopheles* larvae squirmed. Feeding on the pollen of the maize that fell into the water-filled furrows, the larvae grew larger than usual, increasing their likely longevity, and with it their reliability as malaria vectors. At the same time, the high-yielding maize negated the need for the fenced home gardens that farmers traditionally kept between their residences and the fields. Instead, they planted their crops right next to their homes, bringing their bodies within easy flying distance of the *Anopheles*-infested maize. It was this altered agro-ecology, researchers speculate, that triggered the unprecedented malaria epidemic that hit the traditionally malaria-free Ethiopian highlands in 1998–1999.[118]

In the first decade of the new millennium, the rapidly growing Indian economy led to a building boom in Mumbai. Stagnant water collected amid the rubble of construction, while construction workers from across the region introduced new parasites into the area, and malaria began to spread. The city's annual monsoon-related malaria spiked.[119] Over the course of 2006, malaria cases in the city rose by 50 percent.[120] Between June and August 2008, more than fourteen thousand cases of malaria were recorded. "The numbers are huge," the epidemiologist Kishor Harugoli said. He places the blame squarely on the construction boom.[121]

It is not as if all environmental disruptions will set in motion developments that will trigger malaria epidemics. The ecology that sustains the disease varies from place to place. But at least with the building of dams and the logging of forests, the actual transformation itself is obvious: Waters go still. Trees fall down. In the case of what may well turn out to be our biggest environmental disruption of all—the changing global climate due to excess carbon in the atmosphere—the contours of the possible disruptions that will strike are more obscure, and thus their impact on malaria even harder to predict.

Nevertheless, climate-change-induced malaria—unlike the malaria caused by routine industrial practices—has already inspired great alarm in the public mind. "It bites, it kills, it's coming to Essex," read a recent headline in the London newspaper *The Independent*. "Malaria . . . Many researchers believe global warming could bring the disease back."[122] "Climate change brings back malaria," a dispatch from the Italian website ANSA warned. "Italy in firing line."[123] "Malaria goes global as the world gets warmer," Singapore's *Straits Times* headline writers added.[124]

And yet, for all the sensation, climate change is expected to be nothing if not variable: hotter in some places, cooler in others, wetter here and drier there. There is no easy equation between any one factor and malaria transmission. The right climate doesn't mean there'll be the right mosquitoes, or the right parasites, or the right human population.

Even when a global warming–induced change is predictable, its impacts may not be. More rain *could* mean more malaria. Or not, if the rain washes away mosquito larvae, say, or deepens water bodies, allowing them to sustain fish, which would eat the mosquito larvae. More warmth *could* mean more malaria. At higher temperatures, malarial mosquitoes bite more and grow faster. The parasite develops more rapidly inside the mosquito, making it more likely that the mosquito will survive long enough to infect people. But then again, England was at its most malarious during the Little Ice Age, and malaria receded from Europe during a warming period. Other factors outweighed the weather.[125]

Climate experts do widely agree on certain effects of global warming. El Niño, a warm ocean current named after the baby Jesus, usually makes annual visits to the shores of Peru for three to six years in a row, after which cool currents rush in, in an opposite phenomenon named La Niña. The alternating currents influence trade winds, the jet stream, and storm tracks that shape the planet's climate. In northeast Brazil, southern Africa, South Asia, Indonesia, and northern Australia, El Niño years result in droughts; in Peru, Colombia, Ecuador, and Bolivia, intense rain.[126]

We know that El Niño years are correlated with a spike in malaria cases and deaths. In the Kenyan highlands, people have long enjoyed malaria-free lives thousands of feet above sea level, where malarial mosquitoes can scarcely survive. But after heavy, El Niño–induced rains in 1998, mosquitoes invaded the region. To the nonimmune villagers, the ravages of malaria that followed the mosquito bites were inexplicable. "In a crowd of perhaps two dozen people," recounts a *New York Times* reporter who visited the area, "no one could say exactly how malaria was spread or how to prevent it." "If you have experience, maybe you can explain it," one girl said to the reporter.[127]

Hundreds perished. The following year, a three-month outbreak took even more lives. Malaria experts were conflicted over what precisely triggered the violent resurgence of malaria, but for the local clinicians, the answer was obvious. "When you think it will disappear, then it rains again," one said. "There will be more stagnant water, the mosquitoes will hatch, and there will be problems again."[128] Today, fifteen districts in the highlands of Kenya are under constant threat of malaria epidemics, compared to just three in 1988.[129] In Venezuela, El Niño is correlated with a 36 percent increase in malaria's death toll. In Sri Lanka, the risk of malaria epidemics grows by 400 percent when El Niño is in play. In northeast Punjab in India, the risk increases by 500 percent.[130]

In 2006, researchers found the malaria vector *Anopheles arabiensis* on the slopes of Mount Kenya, where the snow cover has started to melt, for the first time ever. Some fifteen thousand tourists visit the Kenyan highlands every year, regularly introducing malaria parasites to the region. Researchers predict that malaria prevalence could rise to 80 percent.[131]

Mathematical models predict other extensions. For example, according to one model, climate change could create amenable conditions for *Anopheles farauti* from the sparsely populated northern tip of Australia to extend four hundred miles southward into Queensland's population centers and tourist hot spots. The last known outbreak of malaria carried by local mosquitoes in Australia occurred in the Northern Territory in 1962. There just aren't enough people living up there to sustain the parasite over time. Not so farther south, where a constant influx of people from malarious Papua New Guinea and the Torres Strait Islands would ensure a robust supply of malaria parasites. Queensland has so far been unable to control other mosquito-transmitted diseases, such as dengue and Ross River fever. If the *Anopheles farauti* mosquito starts passing on malaria parasites in Queensland, the result would be far different from that in the sparsely populated far north.[132]

Now that we know the delicacy of malaria's ecological stability, could we preserve each mosquito-ridden waterway and marsh and microhabitat in pristine stasis, tiptoeing so lightly on the landscape that nary a pebble ripples its glassy surface? It is hard to imagine. Such a nonintrusive existence may once have been possible, during those early days when the global population consisted of a few bands of hungry hunters hauling themselves across the savannah. There's just too many of us now. Even if we surrender our machines, blunt our saws, and fill our mines, the appetites of our hungry, growing population will continue to scar the terrain. We will clear the land, furrow the soil, and stomp into warm mud—and by so doing, risk tearing into malaria's gauzy architecture.

Whether climate change or any of the other ecological disruptions we've set in motion will worsen malaria is just speculation, for now. But one thing is clear. We manipulate the environment as surely as beavers build dams, creating a constant stream of new and altered conditions that mosquitoes and parasites can exploit. When they do, the malarial shock waves spread far and wide.

5. PHARMACOLOGICAL FAILURE

An entire shelf in my medicine cabinet is lined with orange bottles of malaria pills. I get them from my regular physician. I tell her I'm traveling to a malarious region; she writes me a prescription; I get it filled at the local pharmacy; an insurance company pays for most of it; I chip in a few bucks. It's a painless procedure.

Sometimes I get the kind you have to take every day; sometimes the ones you take once a week. Either way, when I visit malarious areas, I pry open the plastic caps and wash down the pills with near-religious conviction. I pack boxes of chocolates, which I store in tropical cupboards, so I can slide the pills into the half-melted sweets for my kids. They actually look forward to their doses.

People vary in their devotion to the drugs, of course. A few, such as an American entomologist I met in Cameroon, are super-careful, like me. When he travels in malaria territory, he rotates between two or three different drugs, and drinks gin with tonic—spiked with antimalarial quinine—just to be extra safe. "I'm a man with a belt *and* suspenders," he says.[1] Others don't bother. I've met as many malariologists who don't take the drugs as do. Terrie Taylor didn't take anything one year, and promptly got infected in Malawi. She was very sick for a few days, she tells me, and remembers lying in front of the fireplace for half an hour, debating the "relative discomfort" of lying on her side versus turning over onto her back. Now she always takes the drugs, because it is just "inconvenient" to be that sick. The worst thing, she says, is that she craved Orange Fanta. "Orange Fanta!" she exclaims.

Some pick and choose, depending on unrelated whims. A poet I know prefers the older drugs, which have the side effect of provoking vivid dreams. He likes the subtly psychotropic effect. "It's good for writing poems," he says.

All of which is to say that antimalarial drugs can prevent malaria as well as treat it, and everyone knows this. All we have to do is open our mouths, and our chosen chemical assassins will seek out and destroy the malaria parasites roosting within us. As prophylactics, antimalarial drugs can be up to 98 percent effective at preventing malaria infection.[2] As treatment, they're crucial. Antimalarial drugs reduce the risk of death from falciparum infection by fifty-fold, sending the parasite levels in the blood plummeting. Even someone sick with life-threatening malaria can reduce their risk of death five-fold with decent antimalaria drugs.[3]

So these days, if a Westerner gets sick from malaria, or dies from it, the question naturally arises, *Why didn't he take the drugs?* The inverse—that the poor die from malaria because they can't get the drugs—we take as a given. But in fact, that isn't really true. Antimalarial drugs are available at local vendors across the malarious world. Synthetic antimalarial medications, available since the mid-twentieth century, can be bought cheaply all over the world. And plants that grow like weeds produce antimalarial compounds, including one that is currently considered a wonder drug. People have known about many of them for almost as long as we've had malaria.

Good drugs—some half as effective, twice as expensive, three times harder to get—have vanquished countless other scourges, from leprosy to rheumatic fever. But despite the range of effective parasite-killing drugs in our arsenal, despite the world being awash in antimalarial medications, malaria flourishes. In the battle between *Homo sapiens* and *Plasmodium*, the parasite is winning. It has eluded attempts to capture it by drug every time.

Our use of medicines to treat malaria dates back to the beginning of human history. Long before the first mosquito bit a human, plants waged a silent, slow-motion war against insects, microbes, and rival plants by producing toxic chemicals.[4] They produce them in their leaves, stems, roots, and barks, and exude them when cut, or attacked from within. The liquid drips and chemical pongs of these silent, primeval green battles play out all around us.

Since the earliest times, we've consumed plants to harness the pharmacological potency of their chemicals (called secondary compounds to distinguish them from those primary compounds that do the work of brute survival). The secondary compounds of plants such as belladonna, poppy, and foxglove—aka atropine, opium, and digitalin—captivate human biochemistry in powerful ways, too.

In the normal course of things, malaria parasites that live inside insects, mammals, and birds don't encounter these botanical bioweapons. But some of those compounds dripping down leaves and branches can do serious damage to malaria parasites. Perhaps it's just a side effect. Then again, 10 percent of the parasite's five thousand proteins retain their algaelike chemistry and remnant chloroplasts. In an earlier life, *Plasmodium* photosynthesized, too.

The shrub mululuza is one of many plants with secondary compounds that provide relief from malaria. Chimpanzees chew on its bitter leaves, as did our African ancestors, suggesting the curious idea that our knowledge of botanical malaria medicines—like malaria itself—may have survived the evolutionary hop from ape to human.[5] Clove, nutmeg, cinnamon, basil, and onion similarly all assuage *Plasmodium*'s appetite, making the body's repair of damage from free radicals—oxygen molecules untethered to hemoglobin—more difficult. This, paradoxically, can help destroy malaria parasites by exposing infected cells to the armies of free radicals that malaria infection unleashes, and may explain why for millennia people sought out and added these nutritionally empty products to their diets.[6]

One of the very best drugs for malaria can be found thousands of miles away from malaria's African cradle, in the bark of the cinchona, a tree that clings to the slopes of the Andes. Like mululuza and basil, cinchona bark started out as a traditional medicine used by the locals for myriad ailments involving fever and chills.[7] The bark teems with complex alkaloids, most likely used by the tree to defend itself against pathogens and herbivores.[8] One of them, quinine, has a striking effect on the *Plasmodium* parasite, interfering with its digestion of hemoglobin. The result: the feasting parasite is poisoned with the undigested toxic residue of its own meal.[9] While it can't prevent malaria infection, quinine circulating in the body can prevent and mitigate illness.

Quinine dropped upon a world beset by malaria like rain on a dry sponge. "Had our bread failed, our wells and the river dried up, we could have endured it," wrote one typical nineteenth-century quinine enthusiast from Michigan. "But to be without cathartic pills and quinine . . . was worse than a bread and water famine."[10] The drug was considered a "sovereign remedy,"[11] a "divine medicine."[12] And when, finally, in the nineteenth century, quinine was made widely available, the Brits were at last able to penetrate malarious Africa. Historians credit the drug, along with filtered water, for Britain's successful 1874 offensive against the Asante Empire in Ghana,[13] ushering in a brief but long-awaited period of European colonization of the African continent.[14]

But despite being a wonder drug for malaria and nearly a half-millennium of usage, quinine barely made a dent in the global burden of malaria.

Jesuit missionaries who'd noticed the bark's effect against the disease in South America introduced it into malaria-plagued European society in the 1630s.[15] Malaria ran rampant in Europe, and cinchona bark was quite possibly the best and most effective medicine ever known. Most of the other medicants then available, such as opium, had a vague, generalized effect on illness. In contrast, cinchona bark could cure malaria effectively and with surgical precision.[16]

It took fifty years for it to gain its rightful stature. Having been introduced by the Jesuits, cinchona fell under a cloud of anti-Catholic sentiment. The Protestant English scoffed at the bark, deriding it as "Jesuit's powder." Elites such as Oliver Cromwell, who successfully led the bloody overthrow of the British monarchy, pointedly eschewed the bark. He died of malaria in 1658, twenty years after the Jesuits brought cinchona to Europe.[17] European elites did, however, approve of a remedy called "Talbor's Wonderful Secret," which everyone, from the son of Louis XIV to the Queen of Spain, extolled. Word got out that the secret remedy got its kick from cinchona, aka the reviled Jesuit's powder, in 1682.[18]

The trouble was, only the tiniest trickle of the stuff could be had. Up until the late nineteenth century, worldwide demand for cinchona mostly had to make do with whatever bark could be stripped off the wild cinchona forests in the far-off and perilous Andes.[19] Political rulers in the region considered the cinchona tree to be their exclusive property. Nevertheless, European entrepreneurs and explorers eager to produce their own lucrative stands of cinchona repeatedly attempted to pirate the seeds out of the Andes, with a nearly comic chain of calamities.[20] The tree and its seedlings, accustomed to steep, forested mountainsides, proved too delicate for long, dank sea voyages across the Atlantic. If it wasn't the heat, piracy, and theft, fires and storms foiled their attempts.

Shortages reigned. Even the most richly endowed enterprises—the American war effort against Britain, the British drive to end the African slave trade—had to ration their quinine. During the American Revolution, the Continental Congress managed to acquire a paltry three hundred pounds of the bark. George Washington's bout of malaria got eight doses of bark, but the regiments of soldiers—including New Milford's Elijah Boardman—had to make do with days of treatment with bowel-purging antimony and mind-dulling opium before their bosses would tap the precious cinchona supply (another reason why Boardman and others sickened on the battlefield and on the trek back home, broadcasting the parasite throughout the land).[21]

The abolitionist British parliamentarian Thomas Fowell Buxton took "every precaution which human ingenuity could suggest" to keep healthy on his 1841 expedition up the Niger River to press African leaders to stanch the slave trade.[22] But his party, too, had to ration quinine. To prevent malaria, they relied mostly on copious volumes of coffee, reserving their quinine supply for a few shots in their wine.[23] (Physicians long suspected that coffee had antimalarial properties, which seemed to explain why coffee-drinking French colonists suffered less malaria than tea-drinking English ones, and may have helped inspire a nation of American tea drinkers to switch allegiances.[24]) Coffee having failed them, disease and death ran so rampant that the expedition was called back home.

A single day's worth of quinine, which French chemists Joseph Caventou and Pierre Pelletier had figured out how to extract from cinchona in 1820,[25] cost about three dollars between 1830 and 1884, which translates to more than sixty 2006 dollars.[26] The moneyed and powerful could at least hoard a few precious grains, but everyday folks had to make do with the cheap knock-offs that littered the market: quinine-spiked pills and potions and chill tonics, which claimed curative powers at a fraction of the price of the real thing. Many of these contained insignificant quantities of quinine, dissolved in copious quantities of alcohol. While not particularly effective at treating malaria, at least they provided plebes with an otherwise verboten drink or two.[27]

Meanwhile, in Bolivia, a cinchona harvester named Manuel Incra Mamani trekked to the hills to collect seeds from a rare stand of cinchona rumored to produce copious quantities of quinine. It took five years to gather the tiny pips. Under threat of death, in 1865 he surreptitiously gave them to a British trader named Charles Ledger, who spirited them out of the country.

The pirated seeds eventually landed in the hands of, among others, the Dutch, who lovingly prepared a bed for them in the rich fertile soil of Java, in present-day Indonesia. The Dutch lavished their horticultural finesse upon the plundered cinchona,[28] clearing thousands of acres of mountainous jungles for the seedlings.[29] But cinchona proved a recalcitrant guest. It preferred the conditions of its Andean home, in areas between three thousand and seven thousand feet above sea level. Closer to the sea, the trees perished; at higher altitudes, they withered to the size of shrubs. Steep mountainsides wouldn't do, either, but only the loose, rich soil of the foothills. Their delicate leaves demanded a constant sprinkle of moisture. Heat greater than eighty-six degrees or cold below forty degrees was intolerable, and a single frost spelled instant death for a young cinchona sapling. The trees would produce not a drop of quinine until they were at least five years old, and they wouldn't countenance the harvesting of their bark until they were at least fifteen years old.[30] It didn't help that wandering Javanese rhinos kept trampling the sprouts.[31]

The Dutch meticulously fulfilled each of these exacting conditions, even as critics condemned the cinchona plantations as "expensive folly."[32] If all went well, the cinchona harvest would be complete within a decade. By the time the trees were age six, their bark coursed with 5 percent quinine,[33] and by age ten, the trees bloomed, their fragrant feathery flowers turning into small fruits teeming with the seeds that would secure the next generation of cinchona trees. Unfortunately, as dedicated as the Dutch were, they hadn't taken into account the miscegenating bees that introduced pollen from worthless cinchona scattered across Java—remnants of earlier attempts at cinchona cultivation—to Ledger's superior, quinine-rich strain.[34] The Dutch had to laboriously deflower thousands of non-Ledger cinchonas around their plantations, forbid farmers from cultivating them ever again,[35] and maintain a rigorously pure seed line, from the original seeds, on the government's estate, for safekeeping.[36] "No tree, not even rubber," mused American horticulturalist Norman Taylor, had "ever had such a long history of patient, intelligent care bestowed upon it."[37]

Establishing cinchona in Java took thirty years.

Overcoming the horticultural barriers to cinchona cultivation proved to be just the first of a series of barriers to widespread quinine use. By 1900, the Dutch produced over five million kilograms of quinine a year.[38] But quinine still did not flow to the masses.

At first the Dutch shared the quinine trade with the Germans, who dominated the factories that processed cinchona bark into quinine. But after World War I, the Allied victors forced the Dutch to stop selling cinchona to the Central Powers and, on the wrong side of the war, the German quinine industry collapsed.[39]

If growing cinchona had been a tad easier, rival powers and enterprising entrepreneurs might have established their own cinchona plantations to compete with Dutch quinine. Cinchona culture being what it was, however, the Dutch lock on the quinine market proved as ironclad as any brand-name patent. And they defended their monopoly with zeal.

To be fair, the Spanish had tried to keep cinchona (and the natural wealth of South America more generally) to themselves, too. During their reign, no one could even go to South America without the permission of the king of Spain, and "nothing on South America could be published," writes the quinine historian Fiammetta Rocco, "without it first being submitted for censorship."[40] (Thus the malaria-plagued Scots in Darién, in 1698, drank whiskey instead of chewing cinchona, despite their dangerous proximity to the bark. They had no idea.) Peru, which gained independence from Spain in 1824, was no more forthcoming. They slapped cinchona on their national emblem and banned the export of cinchona seeds.[41] When Mamani spirited those pirated cinchona seeds out of the Andes, government authorities detained and tortured him for twenty days.[42]

The Dutch were happy to sell the world quinine, but given the economics of cinchona farming, they'd do it only at a suitably high price. After all, if the price of quinine fell, their cinchona farmers in Java would rip up the unprofitable trees and plant tea instead, they said.[43] They set up an agency in Amsterdam called the Kina Bureau, which dictated the terms on which the malarious masses would get their meds.[44] If the price of quinine fell, the Kina Bureau would order cinchona plantations destroyed, quinine held off the market, or bans on the export of planting material.[45]

There wasn't much that national governments could do about this, let alone the average fever patient. In 1927, the U.S. Department of Justice prosecuted the Kina Bureau for violating U.S. antitrust laws,[46] and staged a dramatic raid on a New York warehouse holding five tons of Kina Bureau's quinine, which the authorities seized and locked away at an army base in Brooklyn. A federal grand jury convened to consider the charges against the Bureau[47] and indicted several Dutch leaders.[48] But it was all just so much empty bluster. None of the alleged criminals of the Kina Bureau showed up for any of the proceedings, nor did they attend a showdown meeting U.S. authorities organized at the American embassy in Paris.[49] "Representatives of the Kina Bureau, whose presence was essential to the gayety of the party," a 1934 issue of *Fortune* magazine snarkily noted, "politely declined to eat cake at the American Embassy." The meeting disbanded in humiliation, and the Justice Department agreed to file a "consent decree" with the Kina Bureau, although, as *Fortune* pointed out, "just who did the consenting, and to what, meant very little to anyone."[50] The Kina Bureau remained, as quinine historian M. L. Duran-Reynals wrote, "in absolute command of the situation."[51]

While the Dutch cinchona planters enjoyed 36 percent profit margins, according to a 1934 investigative report in *Fortune* magazine,[52] nine out of ten malaria victims fevered and chilled oblivious to the bitter taste of quinine, their coffers too modest to pay the Kina Bureau's price.[53]

• • •

Lest one suspect the Dutch of some special craftiness, the British failed to adequately distribute the quinine they grew, too.

The Raj grew their own pirated cinchona in colonial India's Nilgiri Hills. Having acquired seeds of relatively quinine-poor cinchona, their plantations never produced much.[54] What little they made, the Brits deposited into small packets and sold to local street vendors and retailers, for resale to the malarious masses.[55] And there, too, needs were not being fulfilled. In 1907, researchers at the Calcutta Medical School found that up to 25 percent of the government-sourced quinine doses available for sale were understrength, most likely adulterated by retailers.[56] Caveat emptor.

In the end, no province in the British Raj ever provided more than 650 milligrams of quinine per person per year, about one third the quantity necessary to treat a single bout of malaria[57]; some, such as the Northwest Province of Punjab, provided just 70 milligrams per person per year. The similarly desiccated private quinine market in India sold about 100 milligrams a year per capita.[58] To take on *Plasmodium*, which killed roughly two million a year in India alone, they'd need a whole lot more quinine than that.[59]

Price, supply, distribution: these weren't the sole obstacles to effective deployment of quinine against malaria. Effective dosing of quinine eluded doctors and patients for centuries. Before the French chemists Pelletier and Caventou figured out how to extract quinine from cinchona bark, people simply consumed the bark itself, scraped off the tree, pulverized, and (sometimes) dissolved in liquid. This practice ensured a highly variable quantity of quinine in the treatment. If it was the right species of cinchona tree, at the right age, there might be some quinine inside the powder. But if it was the wrong species, or the right species but the wrong age, there could be none at all. Unsurprisingly, sometimes the bark worked, sometimes—

as in the case of Britain's Charles II, who died of malaria in 1685, despite taking cinchona—it didn't.[60]

Even after doses with relatively standard quantities of quinine became available, problems with effective dosing lingered. Some clinicians in Europe and the United States felt quinine should be reserved until late in the disease; others argued for tiny doses; still others for bloodletting instead.[61] The recommended prophylactic dose of quinine during the 1850s—based, perhaps, more on economics than on pharmacology—hovered around a measly 2 grains a day, or about 120 milligrams.[62] That's perhaps a third of the dose later proven effective for malaria prevention.[63]

The missionary doctor David Livingstone traveled through Central Africa between 1850 and the 1870s. Not surprisingly, Livingstone found that despite taking the low dose religiously, some of his party fell ill with fever. When Livingstone encountered a comatose Portuguese officer in what is now southern Mozambique, he rakishly tried something different: massive doses of 30 grains, or 1.8 grams. One can only imagine how this might have appeared to his medical colleagues. Reckless, perhaps. In fact, this is roughly the quantity of quinine considered necessary to treat an adult falciparum case today. Livingstone knew he was onto something.

A national celebrity for his exploits in Africa—people followed him around London asking for his autograph—Livingstone had the kind of panache that could grab the attention of the sclerotic medical establishment.[64] Even after heroic quantities of quinine failed to revive his beloved wife, Mary, dying of malaria in a tent pitched along the Zambezi River, he brokenheartedly continued his advocacy for higher doses of the drug.[65] In time, both the British and the U.S. military came to adopt high-dose quinine therapy as standard treatment for their troops.[66]

If the distributors of quinine failed to disseminate enough of the drug, and the scientists and doctors often failed to expertly administer it, it's only fair to point out that the patients failed, too. Too often, people would simply refuse to take the drug, even when it was given freely and at effective doses.

Not for inconsequential reasons. At the appropriate dosage, quinine is not a benign drug. Unpleasant side effects are so common that physicians early on gave them a name: cinchonism. Cinchonism included tinnitus, deafness, headache, nausea, and even visual disturbances. And those were just the expected side effects. (Livingstone recommended taking quinine "until the ears ring.") Quinine could also, albeit rarely, cause severe bleeding, a dramatic decline in white blood cells, abnormal blood clotting, and renal failure, all of which could kill.[67]

There was also the matter of blackwater fever, a mysterious and fatal illness associated with quinine therapy, in which quinine-imbibing patients suffered diarrhea, vomiting, abdominal pain, and finally a jet-black urine and death.[68] Blackwater fever was "perhaps the most important disease, medically and economically, affecting Europeans in the more malarious regions of the tropics," noted the early twentieth-century British malariologist John William Watson Stephens.[69] Although scientists have never established a clear link between quinine and blackwater fever, it's probably not unfair to blame quinine for it—the syndrome died out after long-term prophylactic quinine therapy fell out of favor in the 1950s.[70]

But malaria is no picnic, either. For someone sick with the disease, the side effects of the drug may seem if not trivial then at least tolerable. Still, try as they might, military officers throughout the twentieth century rarely persuaded many of their troops to take the prophylactic quinine as prescribed. In fact, during World War I, soldiers failed to take their prescribed prophylactic doses of quinine so often that the higher-ups began to suspect the men wanted to be sick so they could be evacuated. They called it "malarial defeatism."[71]

A similar reluctance to down the required dose thwarted Italy's ambitious state-run program of free quinine distribution. The idea behind the program, which started in 1902 and continued until World

War I, was to douse the populace with such quantities of quinine that the survival of the parasite itself became untenable. The famed German bacteriologist Robert Koch figured the method could knock out malaria in as little as nine months.[72]

But the "divine medicine" didn't seem so exalted to the locals.[73] For too long, quinine had been a rich man's luxury, and Italian villagers mistrusted the seeming ease with which they could suddenly acquire the drug. "The countryside teemed with dark rumors of a diabolical plot," the historian Frank Snowden writes in his history of malaria in Italy. The government wanted "to rid the nation of its surplus population," some figured; it was a trick to collect more taxes, others speculated.[74]

Despite the huge toll of malaria in Italy and quinine's great promise in fighting it, the majority of Italy's free quinine was never consumed.[75] At least not by *Homo sapiens*. Frightened and skeptical, many peasants fed it to their pigs.[76]

Badly used, misunderstood, and poorly distributed, quinine still prevailed as the world's sole malaria drug (save for traditional remedies used locally) until the 1940s. And so it might have stayed, too, had it not been for the catastrophic quinine disasters of World War II.

At first, malaria was the least of the concerns of American generals prosecuting the war in the Pacific, even as scores of nonimmune troops pushed deep into the malarial jungles and swamps in New Guinea, the Philippines, and elsewhere. The malariologist Paul Russell, envisaging multiple bloody malaria epidemics, visited with American military leaders in New Guinea to plead for greater attention to malaria prevention. They dismissed his concerns out of hand. "If you want to play with mosquitoes in wartime," one told him, "go back to Washington and stop bothering me. I'm busy getting ready to fight the Japs."[77]

Back in Washington, the U.S. government felt prepared. By the time the country entered the war, six million ounces of quinine sulphate had been stockpiled and plans were in place to fortify the U.S.

quinine supply with new shipments of tablets from both the Kina Bureau and cinchona plantations outside the cartel, such as a new one in the U.S.-controlled Philippines. In 1921, the governor-general of the Philippines paid the Dutch four thousand dollars for a bottle of cinchona seeds from their quinine-rich Ledgeriana trees, which he then planted in the southern Philippines, on the island of Mindanao.[78] By 1941, the Mindanao plantations produced around two thousand pounds of quinine a year.[79]

But then, war being war, Germany invaded the Netherlands. One of their first orders of business: hijack Amsterdam's stores of quinine and send them to Berlin.[80] The crushing of quinine's queen bee was ominous enough. Then, in 1942, Japan invaded Indonesia and took control of Java's cinchona plantations (along with its tin, rubber, and other tropical riches).[81] Within a matter of months, 95 percent of the world's quinine had fallen into enemy hands.[82]

As the flow of quinine ebbed, *Plasmodium* flourished among the nonimmune troops. In Papua New Guinea, malaria laid claim to four times more casualties than the ferocious battles themselves, with more than 70 percent of Australian soldiers down with malaria.[83] Every single soldier of the U.S. Americal Division, sent to Guadalcanal in the Solomon Islands in late 1942, came down with malaria, some more than once. According to what General MacArthur told Paul Russell, "one-third of his fighting men were in the throes of malaria, one-third were recovering, and only the final third were truly fit for combat."[84] Overall, malaria sickened 60 percent of Allied troops in Southeast Asia.

Quinine stocks in American shops dried up entirely. "If you require quinine, you must now have a doctor's prescription," *Harper's* magazine complained, "and even with that you will probably get only the ground-bark solution."[85] Malaria cases in India rose precipitously, up to one hundred million by the end of 1942, *The New York Times* reported that year.[86]

But the most notorious epidemic occurred on the island of Bataan, in the Philippines, where American and Filipino troops had fled from a Japanese invasion. With 85 percent of the troops sick with malaria, there wasn't enough quinine for the sick to finish their courses, let alone take the drug prophylactically.[87] Under fire, the weakened, emaciated troops—they'd been eating their own horses by then—surrendered to the Japanese. It was the largest surrender in American and Filipino military history. The Japanese took more than fifteen thousand American and sixty thousand Filipino soldiers as prisoners, forcing them to march one hundred miserable kilometers to prison camps.[88]

It was an outrage, fumed American legislators in Congress.[89] Bataan had been lost "not because the ammunition was gone," as *The New York Times* pointed out, "but because the quinine tablets gave out."[90] The years of reliance on a single highly controlled source for a lifesaving medicine now appeared recklessly foolish. The Kina Bureau had "ruthlessly stifled" competition in the quinine market, a California legislator railed, with "preclusive purchasing, subterranean political activity and indirect economic pressure."[91] Meanwhile, the Japanese, by controlling the Dutch monopoly, "could have all the bark [they] wanted," quinine historian Duran-Reynals pointed out, allowing them to conquer mosquito-ridden China unworried by malaria.[92] To add to their antimalarial superiority, the Japanese army employed elderly women to perform the job of "net tucker-in" to secure the mosquito nets after the soldiers got in their beds.[93]

Allied governments quickly ramped up their efforts to stanch malaria by other means. General MacArthur appointed Paul Russell as chief malariologist for the U.S. military,[94] and Russell assigned malaria control and survey teams to every combat force present in malarious regions—all in all, more than two hundred units, each of about a dozen people. Their supplies and officers leapt from tenth priority in overseas shipment to first.[95] The U.S. Army newspaper reminded troops to keep their sleeves rolled down to avoid mosquito bites. The Armed Forces Radio broadcast so many antimalaria messages that troops called it the Mosquito Network.[96] Even the famed author Theodor Seuss Geisel, aka Dr. Seuss, brought his talents to bear on the malaria problem, producing a cartoon distributed by the U.S. Army in 1943. "This is Ann," Seuss wrote, under a comic rendering of an *Anopheles* mosquito.

> She drinks blood! Ann moves around at night (a real party gal) and she's got a thirst. No whiskey, gin, beer or rum coke for Ann . . . she drinks G.I. blood . . . Never give her a break. She can make you feel like a combination of a forest fire, a January blizzard, and an old dish mop. She will leave you with about as much pep as a sack of wet sand and now and then she can knock you flat for keeps . . . Bathing and swimming at night where Ann hangs out really is asking for trouble. Head nets, rolled-down sleeves, leggings and gloves may seem like sissy stuff and not so comfortable—BUT, a guy out cold from MALARIA is just as stiff as the one who stopped a hunk of steel. Now IF you really are looking for trouble and you don't want to miss [out]—just drop down to the nearest native village some evening. The places are lousy with fat little Anns sitting around waiting for you with their bellies full of germs. They stock up on MALARIA bugs from the home-town boys and gals and when they find a nice new sucker they give him the works.[97]

As the troops digested these morsels of wisdom, scientists back home rushed to screen tens of thousands of compounds in search of a drug that might replace the divine remedy they'd lost. This massive effort at long last ushered in a new chapter in humankind's fight against malaria, with the development of powerful, easy-to-manufacture synthetic chemicals that could lay waste to malarial mosquitoes and the parasites they harbored.

Ironically, given the greater urgency of the Allied nations' search for a quinine replacement, German researchers at I.G. Farben were the first to discover the antimalarial drugs quinacrine and chloroquine, which were only later stumbled upon and developed further by American scientists.

Quinacrine persisted in the blood for a week, which could be useful, but it yellowed the skin and could sometimes trigger psychotic reactions. It also wasn't nearly as effective against vivax malaria as quinine. Quinine-deprived American troops forced to take it during the tail end of the war hated it.[98]

Chloroquine was another matter. Like quinine, it killed malaria parasites by interfering with their ability to metabolize the iron compounds in red blood cells known as heme.[99] But chloroquine departed from its pharmacodynamic cousin in every other way. Where quinine failed, chloroquine excelled. It lasted longer. Its side effects were insignificant. And best of all, it could be churned out in factories as reliably as widgets.

Produced cheaply by drugmakers all over the world, chloroquine exploded into the postwar global marketplace. Elites positioned their bottles of chloroquine prominently on their dinner tables, next to the condiments.[100] Troops assembled for "chloroquine parades."[101] The American drug company that started promoting chloroquine in 1947 claimed it was eight to thirty-two times more effective than quinine,[102] and people, it seems, took them at their word. In Africa, chloroquine overtook aspirin as the drug of choice for fevers, and even aches and pains. This was encouraged by top malaria experts, remembers WHO's José Nájera. "Chloroquine, it was said, should be treated as a commodity, not a drug."[103] Why bother even trying to diagnose malaria before taking chloroquine? The pill, experts advised, should be simply taken as soon as a fever comes on, as "presumptive treatment," by sufferers in their own homes.[104]

The divine medicine quinine, with its quaint botanical heritage, fell into obscurity. The cinchona plantations in Java were left to wither after the war.[105] The German military knocked down the imposing statue erected in Paris commemorating Pelletier and Caventou for their heroic extraction of quinine, melting it down for metal to turn into weapons.[106]

Under siege, the Kina Bureau's tactics grew increasingly unhinged. They secretly paid syndicated newspaper writers to propagandize for quinine. Their collaborators "would not dare write more than eight or nine articles in any one year," however, "in case the syndicate should become suspicious of their featuring quinine too regularly."[107] The bureau promoted risky self-medication with quinine, even though "this might lay us open to attack by the medical authorities," as one Kina Bureau promoter put it.[108] They also tried to engineer criticism of anti-quinine medical authorities. "The best method of attack . . . must appear on the surface to be a spontaneous protest by some outstanding man in this country," one noted, "while it is true that I would supply such a man with all the material." They even toyed, pitiably, with the idea of raising money to rebuild the statue of quinine chemists Pelletier and Caventou.[109] But for nought. In the giddy celebration of chloroquine, the world's hunger for quinine steadily declined.

Plasmodium did not survive for millennia by virtue of some unerring killer instinct, unfailingly homing in on the immunological loopholes and secret hiding places in its prey. It survived by being more adaptable than its hosts. Each species of malaria parasite boasts scores of genetically distinct strains, each with a unique set of strengths and weaknesses. Some parasites may be especially adept at, say, quick reproduction inside the mosquito. Others might be skilled at avoiding capture by immune cells. And the various species hang out together, playing out their rivalries and alliances inside our infected bodies. The human host can, albeit painstakingly, devise a defensive maneuver or two to fight them. But we are single individuals, relatively fixed in our capacities. The parasites infesting us, on the other hand, comprise a rapidly regenerating mini-civilization. Even if some fall prey to our defenses, there will be others who won't—and whose progeny will rapidly conquer the body.

The parasite's tremendous adaptability most likely escaped observers during the quinine era, when the drug's strikes against the parasite remained weak, sparsely distributed, and sporadic. It probably remained hidden during the 1950s, too, as chloroquine consumption took off. But inside drug-dosed bodies all over the world, parasite populations found themselves under assault. Under those conditions, the few hardy individual parasites that could withstand the toll were suddenly plucked from obscurity.

The first signs that malaria parasites could resist synthetic drugs cropped up during the tail end of World War II, as the Allies dosed their troops with the hated quinacrine.

With widespread distrust and dislike of the drug, General MacArthur had instructed the director of medicine for the Australian army, Neil Hamilton Fairley, to provide solid evidence of quinacrine's effectiveness, especially under the wartime conditions in which the drug would be used.

This Fairley did, conducting a series of wrenching human experiments in the unlikely locale of the lush Australian highlands west of Cairns.[110] He rounded up hundreds of volunteers—Jewish refugees and injured soldiers among them—to be purposely infected with malaria and then rigorously dosed with quinacrine, to show how well the drug worked to prevent illness. But he didn't stop there. He exposed the volunteers to hundreds of bites from infected mosquitoes, made them march more than two hundred kilometers in a matter of a few days, with minimal breaks for rest and food, injected them with adrenaline and insulin, and herded them into the freezing chamber of the local meat works. Still, quinacrine quelled their experimentally induced malarias. The drug, in other words, worked.[111] ("They never told us anything," recalls one subject who survived the trials. "At first I didn't realize it was dangerous . . . I thought it would be an adventure and that is why I went," remembered another.[112])

In 1944, Fairley announced the results at a high-level military conference, in Atherton, Queensland. "It has now been amply demonstrated to you that the control of this disease is in your own hands," announced the director general of Australia's Army Health Services.

"It is purely a matter of training and discipline in all antimalarial measures, particularly the taking of Atebrin [the brand name for quinacrine] . . . I have no hesitation in saying, most emphatically, Gentlemen—the ball is now in your court."[113]

A few months later, more than seventeen thousand Australian soldiers landed along the northern coast of New Guinea. There was no ambiguity about what they were to do to avoid malaria. Quinacrine had the stamp of approval from the very highest authorities. It is hard to imagine, given the known malariousness of New Guinea and the weight of confidence in quinacrine, that the troops did not take it religiously. But just to make sure, if they didn't, their platoon commanders were under threat of removal. And it worked. For months, the troops stayed healthful despite the swarms of malarial mosquitoes around them.

But then, about three months into the campaign, the troops started to advance along the coast. Inexplicably, their antimalarial armor of quinacrine seemed to falter. Three hundred and fifty soldiers fell ill with malaria.[114]

News of this must have jolted the troops like a bolt of lightning. Nobody particularly liked taking quinacrine, and these infections could only have been seen as an utter betrayal. The hated drug had failed them! But the leadership, Fairley's faith fresh in their minds, suspected just the opposite. Having worked so hard to implement the distribution of quinacrine, how could they think otherwise? For them it couldn't be the drug that had failed. Rather, the troops had failed to take their medicine.

Amid the controversy over the three hundred and fifty fevered bodies, the troops' commander, Major General J.E.S. Stevens, urged greater compliance. "Brigadier Fairley . . . is the most eminent malariologist in the world," he announced.

> He has devoted his whole life to this disease. He carried out the most extensive investigations and experiments made in the history of mankind and from that made a definite pronouncement that

> 1 tablet of Atebrin per day will suppress malaria. It would be presumptuous for us to suggest that this authority is not absolutely correct. There must be no weakening of our faith in his doctrine, otherwise the whole structure of antimalarial measures will collapse.[115]

To restore the drug's reputation and the protection he believed it provided, Stevens cracked down even more heavily on the wilting troops. Not only did the troops have to line up to receive their tablet, he declared, but the commanding officer was to place the tablet in each mouth himself. After the soldiers drank some water and swallowed, they would be required to call out their names "in a loud voice," and finally, to prove beyond a shadow of a doubt that the pill had in fact gone down their throats, they were to open their mouths for a thorough inspection.[116] In addition, the soldiers were banned from taking off their long-sleeve shirts, trousers, and boots, even "for the purposes of ablutions." Every two hours from dusk to nightfall, whistles were blown, signaling the troops to coat another layer of mosquito repellent onto their unwashed clothes and bodies.[117]

And yet despite the humiliating pill popping and the stinky repellent smearing, malaria rampaged. Every week, another seventy soldiers came down with the disease. The day before a brigadier visited, a falciparum-infected lieutenant colonel became so sick he had to be evacuated.[118]

It must have seemed to officers that the New Guinea troops were in the throes of some kind of willful defiance. The instructions to take the quinacrine had been clear and unequivocal. Such a dangerous breakdown of order would have been shameful enough. But what the higher-ups found, when they arrived to investigate, might have been even worse.

Utterly cowed, the soldiers weren't just taking the drugs, they were wallowing in them, taking even more than required. They took twice the recommended dose. They took it not just to prevent malaria but for aches and pains of all kinds. "One combatant officer said the troops preferred [quinacrine] to aspirin for headaches."[119]

Finally, Fairley arrived on the scene to evacuate some infected troops to Cairns for study. Perhaps the quinacrine was not being absorbed properly, he thought. Or perhaps the troops had been infected by some new, more virulent kind of malaria. Or perhaps they didn't have malaria at all and had been sickened by something else entirely. He'd study their blood, make sure they were getting enough drug, and pinpoint the pathogen that had struck them.

Back in Cairns, Fairley found that malaria parasites had infected at least seven out of nine soldiers. But all of them, incredibly, had quinacrine coursing through their veins. It was a finding, one malariologist later wrote, "without precedent in the history of man's fight against malaria."[120]

The obvious conclusion was that malaria parasites had achieved the ability to circumvent quinacrine's killing mechanism. Perhaps a strain of the parasite had mutated.[121] Most of the soldiers would have had no acquired immunity to *P. falciparum*, so the parasite inside their bodies would have been able to multiply prodigiously, providing ample opportunity for the development of mutations, which occur at a rate of one in a billion.[122] Or perhaps the drug-resistant parasite had been there all along, its hidden abilities unnoticed until quinacrine came along.

Wherever this parasite came from, it rapidly spread, thanks to heavy quinacrine use among not only the Australians but also the similarly malaria-plagued enemy troops, positioned nearby. And the more religiously the soldiers took their quinacrine, the more successful the impervious parasite grew.

But Fairley could not bring himself to face what these findings meant. It was difficult enough getting the troops to take quinacrine without suggesting its abject failure. He still suspected that the troops—not the quinacrine—had played some role in diluting the drug's killing power. After all, only half as many officers got infected as troops. And why would the parasite defang the drug only in New Guinea and nowhere else? So Fairley did his best to protect the reputation of the drug he'd worked so hard to elevate. When he wrote up his results, he allowed that a few parasites circumvented quinacrine, but most of the cases of malaria had occurred, he said, because of failures of "discipline."[123]

To be fair, there was no way for Fairley to have known how this quinacrine-resistant parasite incubating in New Guinea would fare elsewhere in the world, or even over time in New Guinea. It could have been a one-off occurrence, an instance of a strange mutant parasite that would never be seen again. Perhaps it would die out on its own, poisoned by some other mutation that came along with its drug-resisting capacities. Circumventing a drug designed to kill couldn't be easy. Surely the ability placed some kind of compensatory burden on the parasite, the way, say, an Olympic swimmer might not be able to run very well. If the drug-resistant parasite had to compete for blood and mosquito with the usual array of malaria parasites, it would be hobbled, a world-class athlete in a neighborhood street fight. It wouldn't be able to spread far.[124]

Which is why the international community—despite having formed, after World War II, a body to oversee global public health, the World Health Organization—equivocated over reports of malarias that resisted chloroquine, too.

The first hints of malaria's retaliation against chloroquine emerged from Colombia and along the Thai-Cambodia border in 1957, just twelve years after the drug was introduced. One day two geologists who had been working for an oil company in Colombia staggered into a Dallas hospital. They were suffering falciparum malaria. But chloroquine, the king of antimalarial drugs, failed to cure them. Not long after, doctors in Bethesda treated a patient who contracted *P. falciparum* in Thailand. Chloroquine didn't work for this patient, either. "Oh my God, here we go," the malariologist Robert G. Coatney remembers thinking.[125]

Chloroquine kills the malaria parasite by concentrating inside its food vacuoles, where the parasite digests its meal. The parasite attempts to expel the drug, but normally can't do so fast enough. But parasites with a gene called pfmdr1 can spit chloroquine out of their food vacuoles fifty times faster than normal.[126] Such parasites, commented the Centers for Disease Control's Peter Bloland, could "eat chloroquine for lunch."[127] Not just chloroquine. Parasites endowed with pfmdr1 can resist other synthetic antimalarial drugs, including amodiaquine, introduced in the early 1950s,[128] mefloquine, introduced in the mid-1970s,[129] and halofantrine[130] and quinidine,[131] introduced for malaria in the early 1980s.[132]

One by one, the malaria parasite defanged each drug thrown at it. The more effective and widely used the drug, the faster the parasite subjugated it. The antimalarial drug proguanil begat the drug pyrimethamine. The parasite resisted both. Pyrimethamine was combined with a drug called sulfadoxine—a popular combination known as SP, or by its brand name, Fansidar. Parasites resistant to the drug emerged the same year they were introduced in Thailand.[133] Hoffmann–La Roche launched the antimalarial drug mefloquine commercially in 1975. Parasites resistant to the drug emerged a year later.[134]

In 1961, WHO gathered its experts to consider the problem. They agreed that the implications of malaria's resistance were "serious," indeed. But like Fairley, they dragged their feet on sounding an alarm. First, the WHO experts wanted to know all the circumstances under which each "alleged" case of resistance had appeared. They wanted confirmation that the drug in question had, in fact, been appropriately consumed and absorbed. They wanted the allegedly resistant parasite to be captured, and for its passage into an uninfected, nonimmune person to be observed.

As serious as the silent spread of drug-impervious malaria was, the need for discretion, they decided, was even more serious. "It is evident that this and other similar records demand the fullest and strictest investigation," their final report advised, "before reports on them are circulated or published."

After all, malaria's drug resistance had not emerged as a coordinated strike, with all drugs failing everyone everywhere. It was more like guerrilla warfare, with rebellions confined by time and place. The old drugs still worked, at least a bit, for some people in some places some of the time. And so the drug war against malaria continued. Only the most exceptional ecological, demographic, and political circumstances could motivate the development of a new chemical weapon, one that could directly challenge the parasite's insurrection.

As WHO cogitated, conditions in the wet jungles of Southeast Asia deteriorated. There, the chloroquine-resistant falciparum parasite enjoyed a fortuitous and unpredictable stroke of luck. Its edge over other parasites was not restricted to the interior of bodies dosed with chloroquine, however prevalent those may have been. Inside two malaria vectors, the Southeast Asian mosquito, *Anopheles stephensi*, and *Anopheles dirus*, from the western Pacific islands, the drug-resistant parasite was able to develop more effectively than drug-sensitive parasites.

What were the odds? But this was just the edge the drug-resistant parasites needed. Its numbers gained on those of its rivals until its descendants, uniformly endowed with the ability to repel chloroquine, had by the early 1960s become not just a pesky guerrilla tribe but the dominant strain in parts of Southeast Asia and western Oceania.[135]

The unlikely partnership between drug-resistant *P. falciparum* and Southeast Asian malaria mosquitoes happened to coincide with the political and geographical disruptions of the 1959–1975 conflict between communist North Vietnam and U.S.-supported South Vietnam. Just as the drug-resistant parasite achieved its hegemony, hundreds of thousands of unwitting, nonimmune people from the lowlands poured into the malarious foothills of northern Vietnam, resettled by the North Vietnamese government to provide logistical support for the burgeoning war effort. For five to ten years, these people suffered "near-starvation and constant illness," according to Pamela McElwee, an environmental historian of Vietnam,[136] while their malaria-naïve bodies provided ample fodder for the drug-resistant parasite to feast. The Vietnamese jungle soon became the world's premier incubator of drug-resistant malaria.

Had the impending epidemic been confined to the resettled villagers, the military authorities in Vietnam would not necessarily have done much about it. But the exigencies of battle soon turned that malarious jungle into the very lifeline of the war effort. The American bombs started falling in 1965. With the U.S. Navy blocking sea traffic, the North Vietnamese started using the jungle's twelve thousand miles of shaded, mosquito-ridden track as a supply route, sending troops en masse to march through the jungle. They called it the Ho Chi Minh Trail.[137] And it was a deadly one. Exposed to the bites of mosquitoes and fed by the malarious locals, the troops had ample contact with the drug-resistant malaria parasite. After one month-long journey down the trail, only 120 of a regiment of 1,200 Vietcong soldiers were fit to fight. "They had a saying," one Vietnamese battlefield doctor remembers. "'We fear no American imperialists, only malaria.'"[138] The bodies of the people who died on the trail fill twenty-two cemeteries.[139]

Chloroquine-resistant *P. falciparum* infected more than eighteen hundred American soldiers fighting in Vietnam, too. Only twelve died, thanks to the ministrations of that quaint old drug quinine.[140] But the only quinine available to the Vietnamese came from the black markets of Hong Kong. For China's Chairman Mao, overseeing the Cultural Revolution, this meant that reeducation for some of the nation's top scientists would have to wait. Mao launched top secret Project 523, a crash research effort to find a new cure for malaria. "In many cases, we had to 'borrow' [the scientists] for a few days from the hands of the

Red Guards or armed factions," recalled the military scientist Zhou Yiqing. "They were very glad to join the project, too."

The scientists gathered at a Beijing restaurant in May 1967 and hatched their plan to comb through traditional medicines and ancient Chinese medical writings for leads.[141] Tucked inside an ancient medical document called "52 prescriptions," dating from 168 BC, they found descriptions of the medicinal properties of *Artemisia annua*, or sweet wormwood tree,[142] an unpretentious little shrub related to sagebrush and tarragon[143] that will grow, like a weed, in any kind of disturbed environment.[144] Inside *Artemisia* flows a fragile compound called artemisinin, which can kill malaria parasites the way bleach kills microbes, by exerting oxidative stress on their cell membranes.[145] The ancients knew it, too: in AD 340, the physician Ge Hong described how a bitter tea of *Artemisia* provided relief from fever (earlier writers described it as a remedy for hemorrhoids).[146]

Artemisinin was among the first ten compounds of the ancient Chinese *materia medica* screened by the scientists of Project 523. But they didn't heed Ge Hong's instructions. He'd written that the entire fresh plant should be soaked in water and wrung out, and the juice used. This recipe would probably have retained the delicate artemisinin, which doesn't dissolve in water or ether, and is extremely sensitive to heat.[147] Instead, as part of the screening process, most likely the Project 523 scientists used the dried leaves, or applied heat to the plants, destroying the artemisinin.

Not finding any antimalarial activity, they moved on.[148]

Some years passed before the Project 523 scientists looked at *Artemisia* again. This time, they took the recommendations of the ancients more seriously, and by 1972 they had successfully isolated artemisinin.[149] Artemisinin kills malaria parasites faster, with less toxicity and in a completely novel way, than quinine or chloroquine. Even if a malaria parasite could spit out chloroquine, it would still be susceptible to artemisinin's effects.[150] The drug can even kill some of the infective forms of the parasite, the gametocytes, against which quinine and chloroquine do nothing at all.[151] Used by the troops in the final stages of the Vietnam War, artemisinin slashed malaria's death toll by 30 percent.[152]

Project 523's Li Guoqiao, now a professor of traditional Chinese medicine at Guangzhou University, and Zhou Yiqing, now of the Chinese Military Academy of Medical Sciences, are no longer bound by Chairman Mao's edicts on secrecy. Together they've revealed the story of artemisinin's discovery, replete with code names, ancient writings, and raging war, and have captivated the press. The *Far Eastern Economic Review* called the story a "scientific fairy tale."[153]

Somewhat less enchanting is the story of what happened after the drug's discovery: obfuscation, neglect, and misunderstanding that delayed artemisinin's widespread distribution for decades, even as drug-resistant parasites slayed scores.

The first few years of artemisinin's cloistering could hardly be avoided, thanks to the war. The Chinese considered the drug's antimalarial powers a military secret, Li Guoqiao said.[154] The first English-language paper on the drug didn't appear until 1979, seven years after the drug was isolated.

But despite the news of artemisinin's extraordinary powers against the rampaging malaria parasite, the drug still remained obscure outside China and Vietnam.[155] Western scientists frowned upon the unorthodox method of extracting it. A German translator mistranslated the instructions for doing so. Copy editors—suspecting an error, perhaps—suggested deleting mention of it from academic papers.[156] Scientific commentators pointed out that the Chinese scientists had used equipment that was "rudimentary" and "obsolete" by Western standards.[157] WHO refused to approve the drug unless production facilities moved to the United States, a requirement with which the Chinese scientists refused to comply.[158] A chasm of mistrust between Chinese and Western science sucked artemisinin into its vortex.

Between 1980 and 1990, artemisinin drugs slashed China's malaria caseload from two million to ninety thousand.[159] But the

Chinese developers of the drug couldn't interest a world-class drug maker in producing the world-class drug until 1994.[160] That year, drug giant Novartis bought the rights for a pill containing an artemisinin derivative called artemether and another antimalarial drug called lumefantrine, developed by the Chinese team to counter the parasite's ability to develop resistance.[161] (Bombarded by two drugs with different destructive mechanisms, the parasite would be hard pressed to develop resistance to both.) Novartis launched the combined drug five years later, in 1999.[162]

A single course of chloroquine or Fansidar, the cheap old standbys, cost less than twenty-five cents. Aimed at wealthy European travelers on safari and the like, Novartis's artemisinin drug Riamet cost forty-four U.S. dollars per course.[163]

After World War II, killers such as diarrhea and pneumonia steadily weakened their grip on *Homo sapiens* around the globe, taking fewer and fewer lives annually. Meanwhile, malaria's death toll steadily grew.[164] The average annual number of reported malaria cases quadrupled between 1982 and 1997 compared to the period 1962–1981.[165] It wasn't just that clinicians witnessed more malaria cases, as chloroquine treatment at home failed and the sick rushed to clinics, where they could be counted. Long-term surveys showed that the resistant parasites actually killed more people, reversing chloroquine's gains. By the mid-1990s, the mortality rate in Senegal had doubled from that of just ten years before.[166] The risk of death from childhood malaria increased by elevenfold.[167] "In poor countries like ours, children have only one chance," said Fred Binka, an epidemiologist from the University of Ghana. "They struggle just to visit a health service, and if they get the wrong drug the first time, they are then found dead."[168]

After Novartis launched artemether-lumefantrine, another five years passed before the international actors who shape international malaria policies threw their collective weight behind the drug. The

World Health Organization's guidelines on antimalaria drug use are followed by health ministries and malaria clinics across the developing world. In 2001, WHO revised those guidelines, recommending that artemisinin combination therapy (ACT) be the very first drug given in malaria cases.[169] Under pressure, Novartis offered to drop the price of Riamet, with a two-dollar-per-pill version called Coartem.[170]

That wasn't enough, however. Coartem still cost ten to twenty times more than, say, chloroquine. Few if any health systems in malarious countries could afford to make the drug available at their clinics and hospitals without outside support. But the Western funders who might have subsidized the cost balked. According to Dennis Carroll of the U.S. Agency for International Development, Coartem was "not ready for prime time."[171] Better to restrict the use of this expensive combo drug, agreed Rick Steketee of the Centers for Disease Control, and distribute cheap bed nets instead.[172] In a 2003 malaria epidemic in Ethiopia, the United Nations Children's Fund expressly refused to pay for artemisinin combination drugs. There wasn't enough supply on hand, they said, and the new drug therapy would cause confusion.[173] On the Myanmar-Thai border, the agency supplied nearly useless chloroquine instead.[174] In 2003, the international health financing agency the Global Fund to Fight AIDS, Tuberculosis and Malaria allocated three times more funding for chloroquine and another World War II–era drug, sulfadoxine pyrimethamine, than it did for artemisinin combo drugs.[175]

And so, between 1999 and 2004, 95 percent of children with malaria in Africa got treated with the old standby chloroquine. At least half the time, the drug failed to work. Some patients improved slightly, and perhaps felt a bit better, thanks to the drug's fever-reducing effects,[176] but they remained infectious, and were likely to relapse. Others simply failed to recover at all.[177]

This situation, after the world's leading health authorities had clearly sanctioned better drugs, led to a huge and caustic outcry among African and malaria aid physicians. Donors' reluctance was "frankly, very difficult to understand," said Médecins Sans Frontières'

Bernard Pécoul.[178] "I couldn't believe my ears," said Binka. "If a physician went to Burma and prescribed chloroquine, they would be negligent," said Canadian health lawyer Amir Attaran. "When UNICEF does the same," he added archly, "it's called 'international aid.'"[179]

International capital finally started to mobilize for ACT drugs in the years after 2004. The Global Fund agreed to provide funding for health ministries to purchase the drugs,[180] and other drug companies, in partnership with nonprofit aid organizations, started developing alternative ACTs, loosening Novartis's monopoly.[181] The nonprofit Institute of Medicine of the National Academies recommended a new global subsidy, to the tune of $500 million every year, to help pay for ACTs (and future malaria drugs), and international experts started scheduling meetings to explore the possibilities.[182] In late 2008, the Global Fund agreed to bankroll a preliminary $6 million for the subsidy.[183]

When there is a lucrative market to be tapped—say, arthritis sufferers or people with high cholesterol—drug makers can rush a new drug from the lab to the market in under a decade. The system for developing and distributing drugs for public health—to supply to people who earn less than a buck a day—is significantly slower. Between artemisinin's development in 1972 and the international community's support in 2004, more than three decades passed. But by then it was arguably already too late.[184]

Over the decades, while the international community argued and deliberated, the malarious masses grabbed hold of whatever artemisinin they could, and the drug leaked out, in hobbled, diluted forms. Drugmakers seeking to turn a small profit, vendors unable to maintain the drug's exacting conditions, downright criminals—together they all attenuated artemisinin's potency against *Plasmodium.*

In the financial and regulatory vacuum throughout the 1980s and 1990s, an underground market in artemisinin drugs thrived.[185] The French company Sanofi-Aventis and Belgian drug makers Arenco and Dafra earned a snappy income selling artemisinin—untethered by a partnered co-drug—across Africa, to those who could afford it.[186] Dafra earned $13 million a year doing this.[187] Chinese and Vietnamese companies launched a plethora of stand-alone artemisinin drugs, too.[188]

Exposing the malaria parasite to an artemisinin unfortified with another drug dared the parasite to develop resistance. The barely regulated private channels through which stand-alone artemisinin drugs flowed—the sweltering corner shops and street vendors of Africa and Asia—could not ensure the exacting standards that heat- and humidity-sensitive artemisinin requires. Given the unreliability of electricity and refrigeration, even the most well-manufactured drug can deteriorate by the time it appears on vendors' shelves in rural Africa. In a study in Nigeria, nearly half of almost six hundred drugs available on the market—antimalarials, antibacterials, and antituberculosis drugs among them—were found to be substandard. Nearly half of the sulfadoxine-pyrimethamine tablets on sale wouldn't even dissolve in liquid.[189]

Without the oversight of WHO or local health ministries, drugmakers dreamed up their own dosing instructions. Many of the artemisinin drugs they sold came packaged with recommendations for a five-day course—enough to make the patient feel better, but not enough to kill the parasite[190]—turning every fevered patient who took the too-short course into a walking incubator for drug-resistant strains. Most alarming of all, the popularity of these drugs spawned a host of even cheaper, and less effective, copies. In 2004, surveyors reported that more than one third of the artemisinin drugs on sale across Asia were downright fakes, pills filled with minuscule quantities of artemisinin, if any.[191]

The very year that the international community mobilized funds to pay for pricey artemisinin combination drugs, scientists found *Plasmodium* parasites that could trick artemisinin—at least in mice.[192] Just as our own genome spat out multiple mutations when under attack by malaria, the parasite's genome was mutating rapidly under artemisinin's onslaught. And some of the mutated parasites displayed a decreased sensitivity to artemisinin. If ACT drugs still killed these parasites, it wouldn't be for long, the medical journal *Lancet* noted.[193] "Are we losing artemisinin combination therapy already?" the journal asked plaintively.[194]

The answer, according to a quiet WHO meeting held in Phnom Penh in 2007, had to be a hushed and shaky yes. The gathered experts discussed how, in parts of Thailand and Cambodia—where chloroquine-resistant parasites had first emerged—artemisinin combination drugs were failing in up to 30 percent of cases. The implications of this failure rate—that the parasite had broken through the sole effective drug left in the antimalarial arsenal—the group concluded, "constitute a regional and global emergency."[195]

"It will be at least ten years before a drug that good is discovered," bemoaned WHO's malaria program director, Arata Kochi, in 2006. "Basically we're dead."[196]

In an unprecedented show of ire, Kochi had already threatened to publicly name and shame the companies if they kept on selling their second-rate artemisinin, and even to disrupt sales of their other products.[197] Some companies bristled. The old accusations—drug monopoly—wafted in the air. "Just put the facts on the table," one Chinese pharma exec pointed out. "The only ACT provider is Novartis. And now you [can] only use ACTs. What do you think?"[198] Even after the bigger companies complied with WHO's demands, nearly two dozen smaller companies with less to lose from WHO sanctions continued to sell artemisinin drugs.[199]

When the leaders of a small village in rural Myanmar found out that fake artemisinin drugs had led to the death of a local twenty-three-year-old, they ransacked every shop and clinic in the area for its artemisinin supplies. Enraged, they heaped the pills into a towering pile and lit a match. The village gathered for the public bonfire, as the smell of the drug that could have killed malaria filled the air.[200]

The criminals who sold counterfeit artemisinin drugs remain at large. Despite the efforts of Interpol, only a single counterfeit drug trader, in southern China, has been arrested for selling counterfeit artemisinin.[201]

The obstacles to our drug war on malaria are not news. Surmounting them occupies perhaps the biggest chunk of time and resources currently spent holding back malaria's tide. International donors, health ministers, scientists, drugmakers, charities, hospitals, clinicians, community health workers—thousands upon thousands of people on the front lines of the war against malaria devote themselves to getting better drugs to more people. International networks pinpoint the spread of drug-resistant parasites, high-tech labs churn out potential candidates for new malaria drugs, and giant philanthropies work out the details on how to pay for it all. Thousands of experts toil in research labs, clinics, and villages across the globe, burning through hundreds of millions of dollars to do it.

No one can accuse us of lack of diligence in our devotion to the magic-bullet cure, the miracle drug, the wonder pill. And yet, though antimalarial drugs are "one of humanity's most precious and cost-effective public health resources," as the nonprofit Medicines for Malaria Venture put it, it's useful to remember this: even if we somehow got our act together to unleash the full power of our antimalarial drugs upon the malaria parasite, we still wouldn't win.

Even when we've deployed more effective drugs more effectively, the best we've been able to do is blow on malaria's fire. The flames die down, as long as we keep blowing. But as soon as we stop, they leap up, more vigorous than ever before.

A 1930s effort in Panama rigorously treated all villagers infected with parasites, sick or not, with quinine and other antimalarials. But the more effectively treated the locals were, the less adept their bodies became at fighting off malaria parasites. "Continuous drug therapy results in a lowered immunity," the scientists explained, "and therefore a predisposition to heavier parasite rates and more frequent clinical manifestations." The treatment rendered the villagers dangerously vulnerable to the parasites festering in outlying areas, which could contaminate their parasite-free locale. The result, then, would be disastrous. "We fear that more harm than good has been done by this method," scientists reported.[202]

In the early 1980s, government officials dosed the entire, nearly two-million-strong population of Nicaragua with a three-day course of chloroquine and primaquine. Reported malaria cases declined for a few months, but within six months, the caseload shot back up to its pre-drug level, and then went on to increase.[203]

Starting in 2004, Project 523's Li Guoqiao oversaw the administration of artemisinin combination drugs to more than twenty thousand Cambodians about thirty miles outside of Phnom Penh, and to more than forty thousand on Mohéli Island, one of the Comoros Islands off the coast of East Africa. In Cambodia, parasite levels dropped precipitously, but as close to zero as they fell, the drugs could not extinguish every last parasite. Malaria hung on, its powers of regeneration fully intact.[204] On Mohéli, the parasite rate similarly fell, but did not disappear. Even after the vigorous distribution and consumption of thousands of doses, a little over 1 percent of the local people and mosquitoes still harbored the parasite.

The resources required to successfully administer drugs on a mass scale are extraordinary. And yet the best that mass drug administration can do is *almost* put the fire out.[205] Malarious embers smolder on, awaiting their next spark.

6. THE KARMA OF MALARIA

According to studies on risk perception, people are most frightened of unknown risks and least frightened of familiar ones. That's true even when the unknown risk is minimal and the familiar risk is colossal. Take the neurodegenerative cattle ailment Mad Cow Disease. It affected a few cows in Germany in 2000, and 85 percent of the German public considered it a serious threat to public health. In contrast, cars kill three thousand people every day the world over. We continue to drive them, thoughtlessly and with abandon.[1] We take only the most trivial precautions—strapping on a seat belt, perhaps—and even that, to be fair, we do under threat of penalty.

For the people who live with *Plasmodium*, the risk of malaria isn't just numbingly familiar. In their lived experience they know that the overwhelming majority of the parasite's incursions are trivial. Most of the time, carrying the parasite means next to nothing: no fever, no chills, no readily discernible symptoms, especially against a gray backdrop of other, more pressing ailments. It may not even be noticeable. Only seldom does one fall ill. Even then, when illness does occur, ninety-nine out of a hundred times, the fever and chills come and go.

Is it possible that the people who live with malaria consider the parasite the way we consider the sedan in the driveway, as that familiar killer?

This is not an idle question. Most of the ways we've devised to destroy malaria rely upon the committed participation of malaria's victims. It is they who must drain the standing water, swat the mosquitoes, wear the repellent, sleep under the bed nets, go to the clinics, and take the drugs. Their understanding of the disease—what it is, where it comes from, how it can be avoided, whether antimalarial actions are worthwhile—is absolutely crucial, and, in many important ways, as divergent from ours as water is from wine.

The road to Chikwawa, a series of villages in the low valley of the Shire River in southern Malawi, runs down a steep mountainside with dizzying hairpin turns. In every direction tall stalks of corn stand along red dirt tracks. Dusty women with babies strapped to their backs emerge from the green, along with knots of children chewing on corncobs.

When malariologists want to study malaria in its natural habitat, they come here, to Chikwawa. From above, it appears to be one sprawling cornfield, with little clearings of mud brick and thatched-roofed huts scattered within it. Down below, on the rutted tracks, a few bicycles pass by, carrying whole families—man pedaling, woman and child seated behind him—and the occasional pickup crammed with passengers. In the villages—and the small village of Namacha is typical—there's no electricity or running water, not a shard of plastic or a stick of furniture. There's a reed granary in the middle of a clearing, a few skinny cows, towering termite mounds. The dominant sounds are the whirring and clicking of birds and insects, and the rustling of the cornstalks. Aside from the international NGOs whose offices line the main road, such as Family Health International, whose blackened building faces the Kuseli Kumrenji Coffin Workshop, the clanging racket of modernity is inaudible.[2]

The subsistence farmers who live here belong to Malawi's dominant ethnic group, the Chewa. In earlier times, they didn't stay in these villages for long. They weren't nomads, but they'd move now and again, leaving lands fallow after the soil was depleted of nutrients. Today, that is no longer possible. Some five hundred thousand people are crammed into this river valley, and they are stuck. Agribusinesses, such as Illovo Sugar, which sits on the edge of Chikwawa, have bought up much of the surplus land, squeezing Malawi's growing population onto the rest.[3]

Women with stacks of wood piled on their heads and men pushing antique bikes laden with towers of branches labor up the hill out of the valley. White plastic sacks of charcoal line the roadside. Besides growing corn and other crops, the people of Chikwawa cut down the trees to turn into charcoal. As a result, more and more of the soil washes away with the rains that start in December and don't stop until May.[4] The Mwanza, a once narrow and deep tributary of the Shire that runs through Chikwawa, is now flat and shallow. Every year it floods, blanketing the land with mud and leaving Chikwawa's village clearings covered with shallow, wet gulleys, ponds, and puddles filmed over with green scum, from which hatch *Anopheles gambiae* mosquitoes, bestowing the people of Chikwawa with 170 infected bites every year, and malaria that lasts all year round.[5]

At least that's how it seems to me.

A small cadre of medical anthropologists make it their business to study how the people who live with malaria think about the disease, and their findings, although seldom referenced by malaria scientists, historians, or policymakers, are riveting. One of the best ethnographic studies on health beliefs was conducted in 1994 by the University of New Mexico's Deborah Helitzer, who spent eight months living in a Chewa village on the eastern edge of Lake Malombe, about a hundred miles upstream from Chikwawa, interviewing people about malaria. Her findings carry profound implications for humankind's struggle with the parasite.

Perhaps the most important finding concerns the etiology of malaria. For us, malaria is a disease caused by a protozoan parasite transmitted by mosquito. For the Lake Malombe Chewa, malaria—which the locals call *malungo*, and lump together with other malaria-like illnesses—is a disease caused by mosquitoes . . . and spirits and jealousy and hexes and bad weather and hard work and dirty water and rotten food, among other things. "This *malungo* came because I went up the hill to collect firewood," one Chewa man explained to Helitzer. "I went there two times in a day and this was too much work." His fever started the following day.[6]

Anthropologists studying rural people in the Philippines in the late 1990s similarly found a widespread "disbelief" in the mosquito as the vector of malaria. In Gambia, they found people believed that "close association with cattle" or with certain nomadic peoples caused malaria. In parts of Guatemala, people think malaria is brought on by exposure to cold or wet weather, or by drinking unboiled water.[7]

Like intelligent design and other forms of magical thinking, these beliefs are not unrelated to actual shortcomings in the scientific explanations with which they compete. Every time mosquitoes bit Lake Malombe Chewa and they did not fall ill with *malungo*, their disbelief in the mosquito theory of malaria transmission strengthened. Ditto for every time they took an antimalarial drug and it failed to work. If the drug didn't work, this meant that the *malungo* was not caused by mosquitoes.

What these beliefs mean is that while our malaria is an eminently preventable disease, for the Chewa, as for other rural peoples living traditional lives, it is anything but. Malaria is everywhere, caused by everything. It always comes, and for people who live in highly endemic places such as Chikwawa, it usually goes, too—like the seasons, the wind, the tide. A Peace Corps volunteer in Chikwawa told me about a deal he worked out, that allowed local merchants to sell mosquito bed nets for around fifty cents. The nets were popular mostly because they repelled pest mosquitoes, not malaria, he said.

Eventually, people got tired of the bed nets and used them to catch fish, even as *Anopheles gambiae* feasted on their children.[8]

It isn't that the Chewa villagers don't understand that destroying mosquitoes' larval habitats, or sleeping under bed nets, or taking prophylactic drugs, or sealing up their houses, helps prevent malaria. And it isn't that they aren't interested in preventing malaria. There are countless accounts of disease prevention in traditional cultures. In traditional Chewa culture, for examples, cattle and people were housed in such a way as to avoid contact with the tsetse fly, and the sleeping sickness it carried.[9] The Chewa also allowed soot from their cooking fires to blacken their huts to repel pest insects.[10]

No, it isn't that they are not interested in effective fixes. It is that, as with people everywhere, there's little interest in fixes that are time-consuming or temporary, or that promise only—in their minds—marginal efficacy. Even if some *malungo* can be alleviated by people avoiding mosquito bites, they can't possibly avoid exposure to the weather, or to hard work, or to the envy of their neighbors.[11]

Partial acceptance of the science of malaria is only one of the differences between our malaria and theirs. In modernized parts of Africa and Asia, most people may embrace the scientific understanding of malaria but still consider the disease the way Westerners consider a headache or a cold or a bout of flu. That is, as no cause for heroic measures. Those who live under endemic malaria do not think of malaria as a killer disease, a predatory wolf to be violently repelled at all costs. For them, it's more like a stray dog: always around, sometimes annoying, mostly harmless. Among the rural poor, "malaria is perceived . . . as a mild disease," notes the Institute of Medicine[12]; a "relatively minor malady," admits a report from the World Health Organization.[13] In a recent survey conducted by Swiss Tropical Institute epidemiologist June Msechu, nearly 60 percent of Tanzanians said they considered malaria a "normal" problem of life.[14] Similarly, nearly every Indian relative of mine reacted to my writing this book with mild puzzlement, as if I'd announced I was working on a book about bunions.

When the Lake Malombe Chewa fall ill with malaria they don't rush to the clinic and hew to the clinician's every suggestion. They suffer their illness at home. If they decide to take some medication, they'll choose and buy it themselves from the corner shop. Seventy percent of all the antimalarial drugs distributed in Africa are not doled out by doctors and nurses but sold to patients privately by street vendors, market sellers, and other retailers.[15] Buyers take the drug until their symptoms are relieved, and share or save the rest for their next bout.[16]

For the vast majority of malaria cases, medical care is the exception rather than the rule. This means it is difficult for experts to know how much malaria is occurring, let alone to act as stewards over its treatment. It also means that folk wisdom plays an inordinately large role in how expert-devised antimalarial methods are implemented.

Traditional Chewa people, Helitzer found, judge the efficacy of the medicines they buy based on taste. Because it is chalky tasting, they consider aspirin harmless. Because it is bitter, they see chloroquine—which malaria scientists find to be the very picture of a safe and nontoxic drug—as extremely powerful. If they consider their case of malaria to have derived from some everyday factor—a change in the weather, say, or a hard day's work—they treat it with only a mild medication, such as aspirin. If they use chloroquine, they use only a tiny bit, because of its perceived great potency. (And they certainly don't countenance using chloroquine or other antimalarial drugs, many of which are bitter, to *prevent* malaria.)

Such practices help them conserve antimalarial drugs, for even at the clinic, Helitzer found—in 1994, at the time of her study, as now—there are continual shortages of antimalarial drugs and health workers invariably dole out insufficient quantities to each patient. This of course virtually guarantees that malaria parasites will be regularly exposed to sub-therapeutic doses of the drug. Drug-resistant strains will therefore emerge, and antimalarial drugs will become increasingly ineffective. Malaria will persist. Traditional practices thereby reinforce traditional Chewa people's assumptions: drugs don't always work and malaria is a normal part of life.[17]

One important way malaria kills African children is by causing a complication called cerebral malaria. Helitzer found that the Lake Malombe Chewa clearly distinguished this alarming and often fatal *malungo* from regular *malungo*, calling it *malungo wa majini*. They could accurately pinpoint its symptoms, too: convulsions, fever, and body contortions. They knew that immediate attention from an expert was required, and they were willing to pay large sums of money to get it.

But not at local clinics or health departments.

Even as a child's fevers turn to convulsions, many rural families avoid the clinic for as long as they can. *Malungo wa majini* translates literally as "spirit fever,"[18] and for traditional Chewa, the proper treatment for spirit fever is a visit to a local traditional healer,[19] who holds that cerebral malaria, with its strange, otherworldly symptoms, is the work of angry specters and phantoms.

Eighty percent of Africans use traditional medicines for their illnesses. Traditional medicine isn't popular because it is cheaper than allopathic medicine—in Malawi, traditional healers charge up to three times more than the local health clinic for their services[20]—but because the locals consider it more authoritative. Unlike the foreigners and city folk trained in Western-style medical and nursing schools who staff the clinics, traditional healers are rooted in the community. They know their patients, and their patients' ancestors, and their children, and all their kin. For this, their word on health matters can carry much more weight than that of the allopathic clinicians.[21] And they are much more accessible. In traditional Chewa villages, there's one traditional healer for every thirty or so souls.[22]

In contrast, the health clinics around Chikwawa, for example, are few and far between. A sick person in Namacha would have to walk over an hour to the district health office, on the main road, and only one villager there appeared to own a bicycle. Getting to Blantyre, and Queen Elizabeth Hospital, would require a formidable fifty-kilometer-long uphill hike into the highlands. Walking it with a sick child or relative would be near impossible save for the most motivated, and none of the people I met seemed to have much if any cash for a seat on one of the crowded, death-defying minibuses that ply the roads. Only the wealthy can afford to drive in cars—the drive from Chikwawa to Blantyre costs the equivalent of twenty-four dollars in gasoline alone.[23]

In any case, according to a study conducted among villagers in neighboring Tanzania, rural parents consider the quality of care in health clinics to be poor.[24] In a study conducted in Senegal, Senegal Research Institute's Tidiane Ndoye found that local people considered modern clinicians dismissive. They "don't take time to listen," Ndoye was told. They always prescribed the same medicine, over and over again, regardless of differences in patients' temperaments or histories. Because of this perceived rigidity, there was no point in going to see them, Ndoye was told. The sick can just as easily go buy the medicine on their own, from the local vendors, who are generally more flexible about payment than the clinics and hospitals anyway.[25]

After all, most clinics and hospitals are not like the modern malaria research ward at Queen Elizabeth Hospital. Many more are like the one I saw in Douala, Cameroon. It is a "hospital" in name only; on the Saturday when I visited, there were no doctors, no nurses. There is no door to open or window to pull shut. There is just a crumbling concrete structure of several open-air rooms, with a variety of sick people resting in each, their relatives clutching currency and containers of food and drink. It's more like a warehouse for the sick than anything else.

That day, two women sat by an unscreened window holding comatose babies glistening with sweat. Both were sick with malaria; for one, this was her twelfth bout with the illness, one episode for each month of her life. Both had been set up with IV drips. Silently their mothers waited in the darkness. The doctor would arrive in two days, on Monday, at which point it could be too late. Every night, the mosquitoes arrived in the hospital. The babies' arms were rippled with evidence of their bites. This was not a place to escape malaria; it was a place to contract it.

And so, for many rural African families, only after home treatment and traditional healers have failed will they resort to bringing their malarial children to clinics and hospitals. The parents at the malaria ward where Terrie Taylor works exude resignation. Dressed in old T-shirts and wraps, they hover over their children's cribs, exhausted, teary-eyed, and silent, clutching faded pink booklets containing handwritten notes from each clinician who has tended to their children. They rarely ask questions.[26]

According to WHO, cerebral malaria should be treated as rapidly as possible, with an injection of quinine and a shot of anticonvulsants. But by the time cases of cerebral malaria arrive in a clinic it is often too late to reverse the course of the disease. The encounter between patient and clinician is ill-fated from the start. If the result is poor, for the patient and her family, the clinic is seen as the option of last resort, the place where you take your child to die.[27] For the clinicians, battered by the sickest cases, malaria is revealed as a dire villain that must be defied at all costs. Malaria thus takes the lives of many hundreds of thousands of children, and the two worlds, like two unmoored rafts, drift farther apart.[28]

As with HIV and cancer patients, many malaria patients live sufficiently long with their disease to be able to exert political pressure on their leaders. HIV-infected people and cancer patients organize marches, staff phone banks, and write letters demanding more research, more funding, more and better treatments. Malaria sufferers, by and large, do not. Because there is no built-in political constituency for malaria, and because malarious nations are generally poor anyway, their government leaders do not generally allocate much political capital to fighting the disease.

Thus, despite the vibrancy of malaria and its shocking death toll, in many malaria-endemic countries there's little political urgency about it. Health authorities can't help but echo the detachment of their constituencies. In 2006 I visited with the Panamanian government's top scientific advisors on tropical diseases, at the Gorgas Memorial Institute, shortly after the Chepo outbreak. Dr. Jorge Motta, director of the institute, told me on the phone and by e-mail that he was far too busy to meet with me, which seemed understandable, given the circumstances. Malaria had quadrupled in Panama, and drug-resistant parasites were creeping toward the capital, a stone's throw from the towering cargo ships that ferry through its canal. I imagined a crush of epidemiological surveys, labs overflowing with blood samples, calls to be fielded from neighboring towns and cities, reports to be filed, and presentations to be made at international gatherings. Of course he was too busy to meet with me.

But when I visited the institute to interview his less-busy underlings, I found the place the picture of inactivity. Motta graciously ushered me into his spacious office. The large conference table was bare and glistening. Some staff members and visitors sat around it, alternately chatting, reclining, and flipping through magazines. Motta ambled in and out, telling jokes. Later in the day, when I poked my head in a couple of times, the scene remained unchanged. At one point, someone wandered in with a few plates of hot snacks. When I left they were still there shooting the breeze.[29]

Anyone who has worked with health authorities in malaria-endemic countries will recognize the pattern. Noises are made about the urgency of the malaria problem, the travesty of thousands dying from mosquito bites—and then the sleepwalker returns to bed. In West Africa, health ministers hold lavish meetings with celebrities and corporate sponsors to announce new malaria initiatives. Malaria is a serious problem, they say. They will work harder to fight it, they say. Then the electricity cuts out and the lights dim. The air-conditioning whirs to a halt. The ministers drone on with their speeches, and the journalists lean farther back into their chairs. They adjust their weight, tip their chins into their chests, and nod off.

Most clinicians in endemic countries consider malaria a fairly boring field in which to practice. One might guess that among doctors and nurses in Malawi, for example, malaria would have the stature that heart disease or cancer has in the West. But local doctors, even at hospitals such as Queen Elizabeth, where cutting-edge malaria research takes place, misunderstand the basic contours of the disease and its toll on the population. Malaria may be a "mild" illness in many people, but the fact that it makes sufferers more vulnerable to other diseases has been known since the early twentieth century, when British malariologists in Malaysia found that mortality from all causes plummeted after they disrupted malaria transmission. Malaria's broad effect on overall mortality has been a cornerstone of efforts to tackle the disease.[30] And yet, when this point came up during a short lecture at Queen Elizabeth Hospital in 2007, the local doctors professed disbelief. They questioned the concept of a malaria prevention method called intermittent preventive treatment, which has been popular in malaria circles since at least 2004.[31] This was as jarring as it would have been to hear clinicians at, say, Massachusetts General Hospital dispute the burden of heart disease, and say they hadn't heard about the preventive effect of daily aspirin to ward it off.

Even the hardworking nurses in Queen Elizabeth's malaria research ward, who spend their days tending to the torrent of malaria-plagued kids who flow through the ward, seemed to have little interest in the basic facts of malaria transmission. Gathered in a cramped lounge for afternoon tea in 2007, the nurses swung their legs over the edge of a twin bed covered in faded green blankets, which they used as a couch. They seemed bored by my questions in English, which made them go silent. Nevertheless, I asked them where the culprit, *Anopheles gambiae*, came from. After a long pause, one said, with some finality, "the swamp." They all nodded. According to the medical entomologist I later spoke to, *A. gambiae* specialize in Blantyre's clear, sunny puddles.[32]

There's a serious dearth of African clinicians specifically trained in malaria. Western donors have launched special programs to entice more Africans into the field, with scholarships and grants for study at Western universities. Trouble is, once they get the special training, they can get a job anywhere in the world. Few are willing to take the pay cut to go back to malaria territory. To entice foreign-trained Malawian clinicians to practice in Malawi, Queen Elizabeth Hospital must offer starting salaries that dwarf those of its most senior staff. It's a high price to pay, yet still there are rarely enough clinicians to tend all of the hospital's sick.

Most malaria deaths occur well outside the official medical system, so like a whale under the sea, *Plasmodium*'s true reach remains maddeningly elusive. WHO estimates that at least 60 percent of malaria cases in Africa, and 80 percent of malaria deaths, go unreported.[33] Even under the watchful eyes of some of the most highly trained malariologists in the world, malaria rides under the radar. Such a common criminal, near ubiquitous, should be easy to diagnose, but it isn't. The gold standard for diagnosing malaria is by microscopic examination of the blood. This takes time, training, and resources—and it is not easy. Technology that can diagnose malaria more simply has been developed, but it has yet to be widely disseminated. In the meantime, a trained technician must scrutinize a thick film of blood smeared across a slide, and because the parasite may lurk in just a few cells, the technician must hunt for it in one hundred different sections of the slide, adjusting the scope for each one. To pinpoint the parasite species, another, thinner smear of blood must be prepared, so that the microscopist can see the subtle morphological differences that distinguish *P. vivax* from *P. falciparum* from *P. ovale*.[34] Since parasite levels vary over the course of an infection, this must be done several times over several days to accurately establish the fact of an infection.[35] And even this may not be sufficient to catch every infection. Using polymerase chain reaction (PCR), scientists can amplify and discern tiny fragments of parasite DNA. In one study in Senegal, two thirds of children whose blood, under microscopic scrutiny, appeared parasite-free were in fact harboring falciparum parasites as discerned by PCR.[36]

The other problem in collaring malaria is that the innocent look the same as the guilty: the blood of a healthy carrier of malaria parasites is indistinguishable from the blood of a mortally infected one. And so while microscopic diagnosis can show that someone *has* malaria parasites in his body, it can't pinpoint whether that person is sick *from* malaria parasites. Indeed, there's evidence to suggest that even the most experienced clinicians, using both clinical and microscopic diagnoses, mistakenly see malaria when some other pathogen is the true culprit. One out of four patients believed to have died of malaria in Terrie Taylor's malaria ward turns out, upon autopsy, to have no malarial pathology capable of causing death. There are no infected cells sequestered in the brain. The patient had malaria, surely, but died of something else entirely.[37]

So how do clinics without the benefit of well-stocked labs, steady electricity, well-maintained equipment, or trained personnel—some don't even have thermometers—figure out if someone has malaria? The simple answer is that they don't. Given a widespread sense of malaria's ubiquity, and the potentially grave consequences should a bona fide case of falciparum infection go untreated, standard procedure calls for "presumptive diagnosis." That is, if there's a fever, presume malaria and dole out the antimalarial tablets and shots.[38]

And so along with a high level of underreporting, there is a high level of overreporting. Nevertheless, statistics are duly gathered. In the mid-twentieth century, the malariologist Leonard Bruce-Chwatt estimated that roughly one million Africans die of malaria every year. Governments, international agencies, aid organizations, and the news media have basically stuck to that assessment. A team from Oxford University, using risk mapping and analyzing a compilation of studies, reports, and unpublished records, estimated 1.1 million malaria deaths in Africa in 2000. In 2001, WHO estimated 1.1 million malaria deaths worldwide, with 970,000 malaria deaths in Africa.[39]

When, in 2008, WHO adjusted its assessment of malaria cases downward, slashing the figure in half, and reducing its estimate of malaria deaths by more than 20 percent, many experts simply shrugged their shoulders. They knew, as WHO said, that nothing had really changed on the ground. "It's better fudging," said one. But "it's still assumption built on assumption built on assumption." Even a "back-of-the-envelope calculation," a prominent malaria epidemiologist added, would render more accurate numbers.[40]

The retired WHO scientist Socrates Litsios, a hunched, white-haired New Yorker, takes obvious pleasure in describing the antics of his solemn and ponderous former employer. He describes WHO's statistical methods this way: Different WHO programs devote themselves to different diseases, from flu to tuberculosis to malaria. Jockeying for public interest, influence, and funding, and working in relative isolation, each tends to exaggerate the burden of its assigned disease. Finally, someone added up the mortality figures for all the diseases, which resulted in an impossible, implausible sum. Embarrassed, WHO held a meeting and literally doled out the numbers, Litsios says. Eyes gleaming, he imagines the scene: "Okay, measles, you get one million; malaria, you get a million; tuberculosis, a million."[41] He roars with delight.

For outsiders, of course, malaria is not some vague, mild, ignorable illness. It's a killer disease, a scourge of the poor, a travesty in the modern world. That's our outsider's perspective, and we stick to it, disregarding, just as we have for centuries, the actual social experience of those who live with the disease.[42]

In the same way we'd dismiss the justifications of an alcoholic, we dismiss malaria patients' apathy as a symptom of their disease.

After all, malarious communities are isolated—for malaria repels outsiders—and their chronic disease burden leaves them weakened and debilitated. The more malaria they have, the more remote and impoverished they become—and they adapt to this reality. They accept malaria, in other words, because malaria itself has lowered their expectations. That's no reason for us, we figure, to do the same.[43]

We portray malaria in our media as a ferocious disease preying on powerless people. A photograph in *The New York Times* illustrating a story on a new antimalarial drug hatched in Western labs, for example, pictures a Mozambican boy lying on a rough wooden bench and gazing mournfully at the camera. The caption explains his obvious sorrow and lassitude by noting that the child has just learned he has malaria and that the disease kills three thousand African children a day. The boy, the reader is led to understand, has just received a death sentence. In fact, in endemic countries such as Mozambique, people get tested for malaria not because they are worried that they have it, but in the hopes that they do, for that would mean they don't have anything worse. The positive malaria diagnosis the boy received would have been, in fact, a solace.[44]

We attribute the underlying conditions that create the social experience of malaria to a simple lack of money and the things it can buy. Malaria in Africa "is just a cash question, basically," said Martin Hayman, a London lawyer and consultant for malaria-control organizations.[45] Money buys better drugs, for example, so we ship the drugs to Africa, and the problem is solved. And yet, even if the quality of antimalarial drugs were to be improved from 85 percent to 100 percent, the overall effectiveness of malaria treatment could improve by only a single percentage point.[46] That's because, as two German epidemiologists found when they posted observers in local clinics and pharmacies, only 21 percent of people with malaria actually visit health centers. Of these, nearly 70 percent don't have a sufficient history taken, and more than 30 percent don't have their temperature taken. Twenty percent are prescribed the wrong drugs at the wrong doses. Ten percent don't bother buying the drugs, and more than

30 percent don't take the drugs as prescribed. The fact that the drugs are only 85 percent effective accounts for a very small portion of the failure in effective treatment. Even with 85 percent effective drugs, only 3 percent of local people were being effectively treated for malaria. If the drugs were 100 percent effective, the epidemiologists reckoned, the percentage of people effectively treated for malaria would rise only from 3 percent to 4 percent.[47]

We send reporters to the malaria-plagued to demand testimony on their need for Western rescue from the malarial wolf. I witnessed one such exchange, between a BBC reporter and a Cameroonian woman holding her deathly ill child. How would she pay for the hospital visit, the reporter demanded. It was an impossibly rude question, delivered sans preamble, but, both parties knew, it was critical for the central premise of the BBC story. The African mother must be captured on record describing her need for money. The woman's face crumpled. The predicament she found herself in, of course, was much more complicated than cash. Whether she was about to cry or laugh was impossible to tell.

Our outsider's perspective on malaria strikes those we seek to help as incomprehensible. Across the malarious world, medical anthropologist H. Kristian Heggenhougen writes, people profess "puzzlement over the focus on malaria." People who live in poverty and who face myriad life-and-death issues wonder "why outsiders pay such attention and resources on what they see as a minor concern within the range of problems they face every day."[48] They "cannot understand why malaria should be selected for elimination," says Thai social epidemiologist Wijitr Fungladda, "rather than their poor living conditions or any other disease."[49] (So what do they want? *The New York Times*'s Tina Rosenberg cites a survey that asked rural poor people just that. "The first three items," Rosenberg notes, "were a radio, a bicycle and, heartbreakingly, a plastic bucket."[50])

This is nothing new. For centuries, outsiders' sense of malaria as a killer disease has collided with the actual social experience of those who live with it. When missionary doctor David Livingstone steamed down the Shire River to Chikwawa in 1859, he came to help save the Africans from the "kingdom of darkness" in which they lived. Although his explorations in Central Africa were not explicitly for the purpose of disease alleviation—Livingstone hoped to "make an open path for commerce and Christianity"—the notion that Africa required such moral and economic uplift rested upon his conception of the continent as backward and diseased, under siege and in need of external rescue. Livingstone, like other Brits of the time, equated climate with health, and good health with good morals, which led him to believe that the heavy toll of African pathogens on British explorers indicated a malignancy in the land and moral turpitude in the people. They care for "no god except being bewitched," Livingstone complained, and were "inured to bloodshed and murder."[51] By establishing missions across Central Africa, Livingstone would, he believed, light the interior and banish this moral darkness.

Livingstone's long-term survival in Central Africa probably rested on the quinine therapy he pioneered, and the fact that he regularly used a mosquito net and wore heavy boots.[52] (*Anopheles gambiae* are especially attracted to the smell of human feet.) But in keeping with the guiding principles of his work, he chalked it up to his own moral strength and respect for good clean living. "It is our conviction that we owe our escape from the disease . . . to the good diet provided for us by H.M. Government," he wrote to *The Medical Times and Gazette* in 1859. He avoided "imprudent . . . exposure to the sun," and partook of "regular and active exercise."[53] Livingstone's project of enlightening Africa proved wildly popular throughout the English-speaking world. His book, *Missionary Travels and Researches in South Africa*, sold a staggering seventy thousand copies. He was the "hero of the hour," enthused *Harper's* magazine in 1857, "a man whose travels, adventures, and discoveries in the interior of Africa are only excelled by the heroism, philanthropy, and self-sacrifice which he has displayed."[54]

But the central premise of Livingstone's project, by his own experience, was deeply flawed. While nineteenth-century British society projected a dark, diseased continent in need of Christianity's spiritual uplift, Livingstone discovered instead that while African diseases regularly felled his European compatriots, the local peoples who joined his expeditions remained healthful.[55] In Chikwawa, he found abundance and good health: luxuriant stands of cassava, beans, tobacco, pumpkins, okra, and millet tended by vigorously singing villagers. Chikwawa's chief did not plead for help or make threatening or depraved gestures, but warmly welcomed the explorer. "We were not to be alarmed," Livingstone remembers the chief telling him, "of the singing of his people."[56]

Still, the celebrated notion of Western benefaction of civilization, culture, and development upon the malarious African masses continued for decades. "The peoples of Africa south of the Sahara are still in an underdeveloped state so far as degree of civilization and culture," noted WHO's deputy director-general at a 1950 meeting on malaria in equatorial Africa. "With untiring generosity and an unflagging desire for progress," he went on, the "very highly developed countries" would contribute their "cultural and scientific resources," to alleviate Africa's malarial burden.[57]

These attitudes derived from not just a different social experience of malaria and other diseases, of course, but also powerful political and economic interests. The British aimed to stamp out the African slave trade, which, besides being morally repugnant, posed an unwanted competitive threat to underemployed British workers.[58] Britain wanted improved access to Africa's natural resources, and hoped to establish political control, too. When the British denigrated Africans' leaders, healers, and faith as chiefs, witch doctors, and devil worship, respectively, and touted Christian morals as the cure for a diseased continent, they had more than Africans' spiritual uplift and public health in mind.

Today, the economic and political context in which Western philanthropists and aid organizations offer help to the malarious masses has changed dramatically. The West's modern fight against African malaria is aimed not at undermining African governments but at collaborating with them. Our economies still rely upon Africa's natural resources, but our public health offerings are not based on speculative conjecture. Clinical trials have proven that antimalarial drugs, bed nets, and insecticides—unlike, say, the Ten Commandments—effectively alleviate malaria.

And yet, muffled echoes of that earlier dissonance reverberate. The Western clinicians staffing the malaria research ward in Blantyre don't seriously consider what the Malawian women all around them think about any of the proceedings. They can't. The mother of a patient in a mysterious coma, according to Taylor, thinks that the problem with the child is the horrible antituberculosis drugs he was given. The mother of a spaced-out and seriously ill girl thinks her daughter has a bad headache. The clinicians don't make much of this. They do what they think their patients need, despite their charges' palpable skepticism. Their achievement, in lives saved, is orders of magnitude greater than Livingstone's—it took five years for his mission in southern Malawi to convert just a single African[59]—and yet, the one-hundred-fifty-year-old gap between the world of the Western clinicians and that of the rural Africans they seek to help remains.

We want to think of Africans as battling an enemy, malaria, so that we can help them fight this enemy. We come—like Livingstone, with his moral righteousness—bearing the best our society has to offer: our riches and our technology. But the fight outsiders would like to wage against malaria isn't always the same one fought by those who live with the disease.

In 2005, the international financing institution the Global Fund to Fight AIDS, Tuberculosis and Malaria agreed to provide $170 million to African governments to buy artemisinin combination drugs. Novartis had knocked down the price considerably and, expecting a flood of orders, kicked up production. By 2006, the company had manufactured thirty million treatments. But few African governments placed orders.[60] "Everything is on the table!" exclaimed one frustrated Novartis rep. "Everything is there! The nets, the drugs, the money—but the orders aren't coming in! I don't know why!"[61] In the end, despite the available funding, African governments ordered less than half of Novartis's supply,[62] and the company had to destroy millions of the arduously produced tablets, for the heat-sensitive life-saving drug wouldn't keep for long. It was a "waste," one malariologist said sadly, a "tragedy."[63]

7. SCIENTIFIC SOLUTIONS

Everything about the Harvard Malaria Initiative, housed deep inside Harvard University's School of Public Health, conveys a single, resounding message: this is where very important, very well-funded activities occur. The building is towering, majestic, especially in contrast with the narrow, rutted Boston streets that stream traffic around it. Security is thick. To broach the building's cavernous underground parking center, your name has to be on a guest list. And to exit the garage, you have to take the elevator, whose green Up button will remain impassive until it receives a signal from a special ID you must swipe through a sensor. There's more security upstairs, and more IDs, and more swiping of barcodes, to pass through heavy glass doors in order to reach the Harvard Malaria Initiative's labyrinthine realm.

HMI is not just a center of malaria research, but an "epicenter" (as its website boasts), with funding support and corporate partnerships ranging from ExxonMobil to Genzyme. The floors gleam, the walls are lined with elegant blond-wood lockers and doors, and the labs buzz with purpose. Researchers here don't need to budget, and in fact have no idea how much their work costs. "If we did the calculations, we'd probably all be flabbergasted," one says.

The two dozen or so graduate students and researchers who work here meet weekly to share their results, in a conference room warmed by an Oriental rug and stately glass-doored bookcases. A buffet table offers them neatly trimmed sandwiches and fruit salad. The meeting's presentations are graceful and articulate, laced with insider jokes about a Harvard education, and received by colleagues with thoughtful, imaginative questions. The only thing that seems to rattle them is the scrutiny of their mentor, Dyann Wirth, the gray-haired, forbidding molecular biologist who presides over HMI, who subjects them to slow, careful, monotone questioning. The day I was there, a few technical queries from Wirth pushed one young presenter over the edge. She misspoke, caught herself, paused, said something, retracted it, and then looked at her audience and laughed nervously.

This is a happy and well-fed gang, exuding optimism and ambition, the very picture of scientific leadership that Harvard self-consciously cultivates. No doubt each participant hopes to produce the kinds of data that will result in the uncorking of one of the champagne bottles poised at the top of one of the conference room bookcases.[1]

HMI, like a handful of other similarly well-endowed malaria labs scattered across the globe, may seem like the venerable product of centuries of unremitting investment in malaria research, the way that, say, the Human Genome Project or the National Cancer Institute can be seen as the results of long-term investments in research on technology and cancer. Surely, the relentless burden of malaria requires an equally relentless scientific response, and one of the top universities in one of the world's wealthiest countries would, as a matter of course, devote a generous portion of its public health research to a global health priority of malaria's magnitude.

Not so. Political and financial commitment to malaria research has been cyclical, sometimes spiking, often falling. Most of the malaria research centers I've visited look a lot more like the one at the Gorgas Institute in Panama, where malariologists toil in a cramped, dingy, and dimly lit corner of the building, mostly using slides, microscopes, and some glassware, the same tools scientists have been using for over a hundred years. High-tech malaria research centers like HMI are not high points on an upward slanting line; they're crests on a wave, leading a wake of deep troughs.

From its founding, malariology has been a fragile, wayward field, vulnerable to the enthusiasms and disregard of a fickle public. Overzealous researchers announce ballyhooed discoveries that turn out to be mistaken. Obscure, underfinanced scientists make breakthroughs that go all but ignored. Important insights, ones that could establish lasting and fruitful scientific paradigms for the field, are met with public skepticism, disinterest, or both.

Malariology's founding question revolved around etiology. What precisely caused malaria? Folk wisdom held that swamps and miasmas were the culprit, but in the late nineteenth century, the new science of bacteriology emerged, exposing for the first time the tiny world of disease-causing microbes. In 1882, the German bacteriologist Robert Koch found the microbe responsible for tuberculosis, and in 1884, the microbe for cholera. In rapid succession over the coming years, scientists fingered the culprits for a range of pestilences: typhoid, tetanus, plague. Similarly, the thinking went, there must be some microbial pest responsible for the age-old scourge of malaria.[2]

Given the peculiar nature of malaria transmission, discerning the strange series of events leading to illness required interdisciplinary collaboration between naturalists, experimentalists, and clinicians. But an insecure scientific establishment, as status-conscious as a pack of wolves, made such collaborations difficult to sustain. Instead, prestige, resources, and influence flowed to the top dog, whether his story rang true or not. Not surprisingly, there were a few costly dead ends.

The economic impediment imposed by malaria couldn't have been clearer to the leaders of the new republic of Italy, founded in

1871. The "Roman fever" shaped the making of the Italian state just as it had the fall of the empire. First, *Plasmodium* claimed the beloved wife of the nationalist revolutionary Giuseppe Garibaldi. According to the subsequent legend, Garibaldi carried her in his arms across the Roman Campagna as she died of malaria, an act of romantic heroism lovingly recaptured by writers and painters.[3] Then malaria killed the first prime minister of the United Kingdom of Italy, just three months after the new state was declared. With 1,500 of 2,200 railroad workers in Sicily sick with malaria, 10,000 of the standing army of 180,000 in hospital with fever,[4] and the new Italian state hemorrhaging millions of its strongest and most hardworking young men to the Americas, Italian railroads, mining companies, and philanthropists begged Italian scientists to find a solution to the problem.[5]

And so they did. One day in the late 1870s, two pathologists, Corrado Tommasi-Crudeli and Edwin Klebs, collected air and mud samples from the Roman Campagna. From the samples, they isolated ten-micromillimeter-long rods, which from the vantage point of their crude microscopes, seemed to develop into long threads. When injected into lab rabbits, the long threads soon had the bunnies heaving with chills and fever. Inside their slaughtered bodies, the pathologists found the ten-micromillimeter-long rods, once again.

The two scientists decided that they'd found the microbe responsible for malaria. It was a germ, it lived in the soil and the air, and they called it *Bacillus malariae.* They announced their findings in 1879.

The scientific method is not infallible, of course, and such mistakes are made, even when the entire economy of a newly formed nation depends on the results.

Counterevidence soon emerged.

In November 1880, Alphonse Laveran, a French surgeon stationed in Constantine, Algeria, peered at a crimson blob on a glass slide. How he found what he did is a bit of a mystery. Most nineteenth-century microscopists soaked their slides in chemicals, their cutting-edge techniques thus unknowingly killing the malaria parasites in their samples and rendering them all but invisible amid the scattered debris of the magnified blood. Those who did examine blood from malaria victims while still fresh, as Laveran did, presumably did so more promptly than he did on this particular day. The blood was still warm when Laveran excused himself from its notice. What precisely he did upon abandoning his slide nobody knows, but whatever it was, it took about fifteen minutes. Maybe it was a cup of coffee.[6]

In any case, during the lull, the drop of malarial blood on the glass cooled. The change in temperature roused the parasites in the sample, which now considered that they had left the warm-blooded human for the cool environs of a mosquito body. Male forms of the parasite would soon be called upon to fertilize female ones, and each started to sprout long flagella and wave them about, in lascivious preparation. Laveran returned to his microscope expecting yet another static scene. Instead, the shocked surgeon caught sight of tiny spheres propelling themselves with fine, transparent filaments, wrigglingly alive.[7]

For the first time ever, *Plasmodium* had been spied by the human eye. Laveran found the creature again and again in his malarial patients, and watched it disappear from their blood after he administered quinine. He knew this was no bacillus, but something else entirely, and he would soon set out to show the world his new discovery, *Bacillus malariae* or no.[8]

Meanwhile, on the Mississippi River Delta, an American army surgeon named George Miller Sternberg attempted to replicate Klebs and Tomassi-Crudeli's results with *Bacillus malariae.* He collected samples from a notoriously malarial area just as Klebs and Tommasi-Crudeli had done. Back in the lab, he isolated the tiny rods and injected them into rabbits. But the bunnies' heaving fevers didn't look like malaria to Sternberg. On a hunch, he decided to inject some other rabbits with an alternative substance to compare their fates. Given an ongoing debate on the infectious capacities of spit, Sternberg decided to use his own saliva. He injected the spit into the bunnies, and it made them sick, too, in the same way as the tiny rods had. So now Sternberg knew that either malaria was caused by both the malaria germ *and his own slobber* or the erstwhile malarial germ did not exist.[9]

Laveran went home to Europe to explain his findings, but the discovery of *Bacillus malariae* had already been adopted by European scientists, business leaders, and the press. "It is a germ," proclaimed *The New York Times*, "and is carried by the mist." After all, the notion of a *Bacillus malariae* fit well with then-current miasmatic theory. *Bacillus malariae* found favor, too, among developers eager to drain marginal wetlands and swamps, an activity they could now claim "ventilates the earth . . . and stop[s] the growth of fever germs." What was not to like?[10]

Laveran was met with stern disapproval. In Paris, France's leading malaria authority, Léon Colin, scoffed at the obscure army surgeon's misinterpretation of white blood cells.[11] In Rome, Tommasi-Crudeli proclaimed that Laveran's creature was just some dead bacilli. The world's foremost microbiologist, German scientist Robert Koch, similarly rejected Laveran's claims,[12] to which the French army surgeon soon appended yet another wild and unheeded speculation: that his malarial creatures were transmitted by mosquitoes.[13]

Laveran's hunch was likely inspired by the work of Scottish physician Patrick Manson, who had discovered, around the same time, the role of the mosquito in transmitting filarial worms and causing the disease known as filariasis. The worms block the flow of the lymph glands, creating huge disfiguring tumors, sometimes as heavy as fifty pounds, particularly on the legs and groin. The disease was endemic in south China, where Manson was stationed with the Imperial Chinese Maritime Customs Service. Neither Chinese nor Western medicine had any treatment to offer filariasis victims, but Manson, unlike his Chinese colleagues, who considered invasive procedures anathema, was willing to lop off the tumors themselves. His amputations freed filariasis patients from their burdens, at least until the tumors grew back.[14]

After scientists found filarial worms in the blood and urine of filariasis patients in 1874, Manson got to thinking about how the worms survived from generation to generation. He knew filariasis was not contagious—the worms didn't spew out in coughs and sneezes, or on dirty hands—and yet they had to have some method of leaving the human body and infecting a new one. He started examining blood samples from his patients, which led him to the discovery that the worms presented themselves in the bloodstream only occasionally, and then for only a brief amount of time. The bloodstream, he surmised, must be their escape avenue. Something was coming to whisk them away. Manson deduced, from this, that the worms must be abetted by some kind of bloodsucker. Filariasis was unknown in England, which ruled out fleas, lice, and leeches, which lived "pretty well all over the world." The vector could, however, be mosquitoes, Manson thought. According to him, mosquitoes were "confined to a limited area of the earth's surface."

Manson built a special "mosquito house," in which he enticed one of his filariasis patients to sleep at night while mosquitoes fed upon the man's exposed body. In the morning, Manson's servant collected the blood-filled mosquitoes resting on the walls of the house and dropped them into glass vials for Manson to examine. Manson sliced the bugs open, exposing the young, sausage-shaped worms within and establishing the mosquito as the vector.[15]

Mosquitoes had been suspected as diseased since early times, in India and parts of Africa as well as in the West. As the first-century Roman agricultural writer Lucius Columella wrote,

> during the heat a marsh throws up a noxious poison and breeds animals armed with aggressive little stingers, which fly upon us in very thick swarms . . . whereby hidden diseases are often contracted, the causes of which not even the physicians can ascertain.[16]

But Manson had read in a natural history book that mosquitoes took but a single blood meal in their lives, before promptly laying eggs and perishing in the water.[17] Since a mosquito that bites only once can't, by its own agency, properly transmit anything, Manson dreamed up a possible modus operandi. The microbes lived in the mosquito, which laid its eggs in the water and then promptly drowned, as per the natural history book's authority. As the mosquito corpse floated along, slowly decomposing, Manson figured, its microbes seeped into the water. At some point, an unsuspecting human ingested the contaminated water. And the ingested microbes infected the human, who would later be bitten by another mosquito, completing the cycle.[18]

In fact, mosquitoes take several blood meals, which is why they are reliable carriers for pathogens—they can both pick up and pass on microbes. Manson's habit of feeding his lab mosquitoes only once reinforced his faulty impression, for they died soon after that single feeding. They probably starved to death.

Manson didn't have much experimental evidence to offer in support of his theory. He'd performed just a single—and illegal—dissection of a filariasis victim in the same hot, windowless room in which the man had died. The "light was very bad and the heat overpowering," he admitted later. The doomed autopsy proved inconclusive. According to the Manson biographer Douglas M. Haynes, conviction drove his theory more than deduction.

Collaboration with scientists with better experimental skills and insights into mosquito behavior could have improved Manson's groundbreaking theory. But rather than seek out collaborators, Manson sought turf. He wanted to stake his claim to the theory, promoted fully intact. Technically, his Scottish medical degree held the same status as one from an English university, but in reality, he was all but shut out of the lucrative London medical scene, and he knew it. He'd need a powerful mentor, someone with say-so among the British scientific elites. There were two top filariasis specialists in the British Empire at the time: Thomas Spencer Cobbold, a fellow of the elite Royal Society, based in London, and Timothy Lewis, stationed in the backwaters of the Indian colonies.

Manson appealed to the authority of the heralded London-based Cobbold. "I live in an out-of-the-world place, away from libraries, and out of the run of what is going on," he wrote to Cobbold deferentially. "I do not know very well the value of my work, or if it has been done before, or better." Cobbold, in the midst of a nasty fight with Lewis over the primacy of his own findings, rushed Manson's fantastic tale to *The Lancet*, establishing himself—with Manson in tow—as the first to discover the disease-carrying capabilities of the mosquito.

Together the two swatted away any inconvenient dissent or counterevidence. Lewis, for example, couldn't find the embryonic and larval worms that Manson claimed to have found, or in any way confirm that the worm was transmitted into water through the mosquito's body. Instead of reexamining their theory, Manson and Cobbold attacked Lewis for his "hesitation and scientific caution." By 1883, "medical and scientific circles in England and on the Continent of Europe" lauded the dead-mosquito-water theory and called the man who came up with it a "keen investigator and accurate observer," as the *British Medical Journal* noted.[19]

The vast majority of people at risk of malaria infection did not have the resources, nor their political leaders the will, to act upon the implications of the latest scientific discoveries. And so these nineteenth-century scientific missteps wouldn't have made much difference in the grand scheme of things, except that, just then, the French diplomat Ferdinand de Lesseps was planning to build a canal through Panama.

Unlike most of the malarious masses around the globe, De Lesseps had both the resources and the insight to implement the most cutting-edge medical discoveries of the day. He was a famous developer who had already successfully built the Suez Canal. And he knew malaria would pose an obstacle to his new project in Panama. The geographers he sent to survey for the new canal had already suffered waves of dengue, yellow fever, and malaria, which they called Chagres fever, after the river. "The white men withered as cut plants in the sun," recalled one visitor.[20] Cholera epidemics had similarly roiled De Lesseps's workers during the building of the Suez Canal, and De Lesseps himself had suffered a horrifying week during which disease killed both his wife and son.

And so, before luring some thirteen thousand workers from Jamaica, Colombia, Venezuela, and Cuba to converge upon malarious Panama, De Lesseps would do all he could to prevent epidemics of disease. He built beautiful hospitals to care for the sick, designed to minimize malarial influences as the leading scientists in Europe described them. Patients' every comfort was considered. Lush gardens surrounded the stately buildings, and scented tropical air wafted freely throughout the wards. The hospital in Colón jutted out over the sea, so that patients could breathe in the fresh sea air. And the legs of each bed were set in small pots of water to repel ants and spiders.[21]

Soon enough, with tropical downpours washing over the machete-wielding workers, "devoured by mosquitoes" every night (as the painter Paul Gauguin put it, before fleeing for Martinique), three quarters of the hospitals were full of fevered malaria patients.[22] "Is M. de Lesseps a Canal Digger or a Grave Digger?" *Harper's Weekly* wondered, publishing an illustration of the rotund De Lesseps despondently shoveling a ditch.[23]

Harper's charged that the canal company was suppressing the causes of death. But in fact there was no reason for De Lesseps's managers to suspect that the mosquitoes that rose from the canal site and from his well-watered hospital gardens and pots of standing water had anything to do with the workers trembling in their beds, or that his lovely hospitals themselves had become purveyors of disease. The canal company's chief surgeon routinely examined the blood of patients in the canal hospitals and found the offending

Bacillus malariae.[24] American medical consultants called upon to examine progress at the canal concurred.[25]

By 1889, malaria and yellow fever had killed an estimated twenty-two thousand people working on the canal.[26] De Lesseps's company went bankrupt, abandoning the muddy, corpse-littered gash. In France, enraged investors blamed human, not entomological, malfeasance: the French government collapsed, and three former premiers, two former ministers, two senators, and a hundred deputies and former deputies were accused of corruption. De Lesseps's son, among others, was tried for bribery.[27]

Meanwhile, the true nature of the mosquito's role in malaria transmission had grown even more obscure. Back in London and hoping to establish himself as a malaria expert—London teemed with malarious Brits returned home from the colonies, needy of medical attention—Manson had reprised his celebrated mosquito theory, but this time to explain malaria. "As the problem and conditions are the same for both organisms," he wrote in 1894, "the solution to the problem may also be the same." That is, unlucky humans contracted not only filariasis but also malaria by drinking water or inhaling dust contaminated with the corpses of infected mosquitoes.[28]

There were several obvious flaws in Manson's theory. For one thing, it required that the malaria parasite survive in the environment on its own for days, or even weeks and months. Scientists such as Laveran—whose discovery of *Plasmodium* had by then been found acceptable—knew that the parasite was too delicate to survive anywhere independently of a host, whether in water or dust.[29] And Italian scientists had established, by feeding swamp water to patients, that water from even the most malarial regions did not infect humans with anything. But Manson stood by his story. Obviously, "the Frenchies and Italians will pooh pooh it at first," he wrote in a letter, "then adopt it, and then claim it as their own."[30]

Debilitated by periodic attacks of gout, Manson himself could not travel to malarial regions to collect mosquitoes and prove the veracity of his theory. He needed an assistant.

Ronald Ross was like many other docs in the British Raj's Indian Medical Service. He professed no particular interest in public health or medicine, or even India, and wasn't especially accomplished. He'd studied medicine mostly because his father, an army surgeon stationed in India, had expected him to. He hadn't done well on his exams, and so qualified only for an Indian Medical Service job in the less desirable Madras branch. But he didn't really mind. There was steady pay to be had, plenty of sport, and only a few hours of work to be done every day.[31] Ross had literary ambitions and spent his days in pleasant distraction, playing sports and writing poetry, while the bloody revolts and epidemics of the Raj swirled around him. He published his first novel in 1889, about the life of a child marooned on an island "not knowing the face of a human being, not knowing the speech of men, nor even their very gestures."[32]

A different kind of hidden world was revealed to Ross when he took a course on bacteriology while on furlough from India. Hunched over his microscope, Ross could spend hours lost in a cryptic world full of shapes and shadows and pulsing with cryptic meaning.[33] Like every other nineteenth-century microscopist, he didn't know half of what he was looking at. But he meticulously described the "clusters or chains of very faint, sometimes brownish, little globules," the "faintest possible indications of a matrix," "delicate, bluish bodies," "bubbles of a pale, yellow, luminous substance," the "flower-like" appearance of "beautiful structures," and "small, beautifully delicate red corpuscles" he saw through the lens. The mysterious microcosm inside a drop of blood smeared onto a fragile glass slide enchanted the young aesthete. Ross was hooked.[34]

Ross published his microscopic meditations, one of which caught the attention of Patrick Manson, who soon had Ross under his wing, promising fame and fortune should he produce evidence in support of Manson's mosquito theory of malaria. "If you succeed in this you will go up like a shot and get any facilities you may ask for," Manson told him. "Look on it as a Holy Grail and yourself as a Sir Galahad."

Thus began one of the most famous and fraught scientific collaborations in medical history. Via a long and well-preserved written correspondence, Manson encouraged Ross to feed mosquito-contaminated water and sediment to healthy volunteers. "The mosquito water, or mosquito dust, should be taken or inhaled first thing in the morning and on an empty stomach," he counseled. "A positive result to such an experiment would be irrefutable."

But the local Indians mistrusted him, Ross complained repeatedly to Manson. "My two cases ran away," he told Manson in 1895, "because I pricked their fingers, in spite of my giving them a rupee a prick!"[35] "The bazaar people won't come to me even though I offer what is enormous payment to them . . . 2 & 3 rupees for a single finger prick . . . they think it is witchcraft. Even my own gardener refused to allow me to prick her finger more than once," he wrote. The mother of a malarial child "fight[s] me about pricking its heathenish little black toe," he complained. Finally he resorted to deception, wickedly informing one reluctant volunteer that the mosquitoes Ross allowed to feed upon his skin performed the therapeutic function of *removing* disease.

Ross's efforts to rear mosquitoes similarly floundered. Like Manson, he knew precious little about mosquitoes. And he didn't have much regard for those who did—"mere naturalists," he called them, "fit for nothing but classifications, making pretty pictures & belonging to societies," he scoffed. He captured and raised the wrong species, whose body was like poison to the parasite. He killed his larvae by placing their bottles in the sun. He couldn't entice adult mosquitoes to bite. They were "as obstinate as mules," he complained. The whole business was "vexatious." Ross once did suspect that the species of mosquito might play some role in malaria transmission, since he frequently found mosquito-ridden places that were malaria free, but Manson set him straight. "The reason for the absence of malaria in certain mosquito-haunted places does not lie in the inefficiency of the mosquito," Manson advised, "but in the presence of something inimical to the plasmodium in it."

Meanwhile, supported by both state and industry, Italian malariologists were hot on malaria's trail. The Italians differed fundamentally from Ross and Manson in their approach to the disease. They conducted interdisciplinary research and embraced the insights of naturalists and of malaria sufferers themselves. While Ross described the malarious Indians he experimented upon as "people who love filth" and "really nearer a monkey than a man,"[36] the Italian pathologist Amico Bignami considered malaria victims "much better informed about malaria than some medical men." Visiting malarious areas, he asked locals about their experiences with the parasite, gleaning important clues about its secret ways. "Many precautions which they take against the fever are taken, one would say, to defend them from the sting of insects," Bignami noted. They avoided going out at night and sleeping outdoors. They closed their leaky windows against insects but not the night breeze, and took "great care of their mosquito curtain," which they wrapped themselves in every night, regardless of the heat.

All of this led Bignami to hypothesize that it was the bite of the mosquito—not its water or dust or air—that carried *Plasmodium* to its final destination. Bignami and his colleagues backed up their hypothesis with experimentation. They fed volunteers marsh water, let them inhale dust from malarious areas, and injected them with blood from malaria victims. The water and dust did nothing, but the inoculation indeed sickened volunteers.[37]

Bignami's team had made progress on the species of mosquito, too. In 1895, the zoologist Giovanni Battista Grassi had joined Bignami in Rome. Grassi, a highly regarded evolutionary biologist, had described malaria in owls, pigeons, and sparrows, among other birds, finding that each bird species boasted its own unique malaria parasite species. For Grassi, as for any naturalist, the species of the mosquito was as crucial a factor in human malaria transmission as the species of the host and of the pathogen itself. Grassi had already thoroughly described the distribution of mosquitoes in Italy (there were at least fifty species), and of the six species that frequented malarial areas, he had narrowed the suspects to just three: two *Culex* species and the true *Anopheles* culprit.[38]

In 1896, Bignami published his notion that the bite of the mosquito transmitted malaria to humans, and with Grassi, he started experimental work to prove it.[39]

Manson and Ross, neither of whom ever achieved much financial success, constantly fretted over securing a living. Manson's tactic was to try to shame the British government into supporting his and Ross's work, appealing to its sense of national pride. "It is little to our credit," he told a medical society gathering in 1894, "that continental nations, whose stake in tropical countries is infinitely smaller than ours, are nevertheless just as infinitely ahead of us in this matter." Ross agreed: it would be so "annoying" if "the Italians do come in first!"[40]

Manson and Ross fought the Italian researchers' findings tooth and nail and clung to their misguided hypotheses to the very end. After all, their very livelihoods were at stake. "Bignami is a pure villain," Ross raged.

> He wants to secrete a mosquito theory of his own . . . He wants to bite into the heart of your theory, suck its juices & then bloat & swell into a discoverer—or rather until he is thought to be one . . . He is quite capable of spreading his six legs over your work & calling it all his own . . . if you have not squelched him already you ought to do it.[41]

Malariology averted the dead end that Ross and Manson urgently steered it into when the two scientists happened upon the same conclusion as Bignami and Grassi, albeit less via methodical inquiry than through serendipity. First, Ross inadvertently discovered *Anopheles* mosquitoes, hidden deep inside a forest (a local servant pointed them out to him). Because he couldn't entice any human volunteers—"for several reasons hospital patients . . . are not convenient to work with," he wrote, "they expect treatment & the papers might talk"[42]—Ross shifted from humans and their malaria to the more easily captured birds and theirs.[43] And while trying to prove Manson's mosquito-water theory, he encountered a strange, delicate structure inside the torn-off head of a mosquito. "This proved to be a long branching gland of some sort, looking like a coil of large intestine," he wrote. "I noticed at once that the rods"—sporozoites—"were swarming here & were even *pouring out* from somewhere in streams." What was this quivering coil, shuddering with tiny squirming parasites? "I still experience, however, the greatest difficulty in dissecting out the gland itself," Ross wrote.

> It appears to lie in front of the thorax close to the head, but breaks so easily in the dissection that I cannot locate it properly. In the second mosquito however there was no doubt, as shown by evident attachment that the duct led straight into the headpiece, probably into the mouth.
>
> In other words it is a thousand to one, it is a *salivary gland*.[44]

In other words, Bignami had been right all along. But Ross and Manson didn't see it that way. In London, Manson promptly declared that he and Ross had solved the mystery of malarial transmission, presenting Ross's work on bird malaria to a meeting of the British Medical Association, enlivening his report with dramatic flourishes such as the reading out of a telegram from Ross on his latest results.[45] "The fat is thoroughly in the fire," Manson reported proudly to Ross afterward.[46] "You will be lionized when you get home."[47] "Well, I have become unbearable with conceit," Ross wrote back. "That was a grand charge! I brag openly about it!"[48]

Of course, Ross had shown that mosquitoes transmit malaria to *birds*, not humans. Bignami had come out with the correct hypothesis first, and Grassi was the first to experimentally infect a human volunteer with malaria through the bite of an infected mosquito, a result published in 1898. Ross had described some gray, dapple-winged mosquito, while Grassi had fingered *Anopheles* specifically. The month before Manson's grand announcement, German bacteriologist Koch announced that *he* had discovered malaria's mosquito vector.[49]

That is, Ross and Manson's stake to the nineteenth century's scientific Holy Grail was as assured as one of those tattered flags flapping on Mount Everest. In the august pages of Britain's leading medical journal, Bignami and Grassi politely cited Manson and Ross and their "interesting observations" on birds.[50] The head of the British Medical Association openly said that the Italians had made the major discoveries in malaria's transmission. When Ross actually went to Italy to visit with Italian researchers, the local papers downgraded him to "an engineer."[51]

The Indian Medical Service, which had repeatedly disrupted Ross's studies, yawned in the face of his fabulous discovery. It forbade him to publish his findings on malaria in birds, and even when it finally gave him time off to complete his investigations, it demanded that he look into other diseases as well as malaria. When he made his big breakthrough, the IMS "congratulated me politely," he wrote, "but they asked no questions, never sent for me, never ordered anyone to inspect or verify my work, never gave me any assistance."[52]

Back in Britain, Ross struggled to secure a well-paying position.[53] He went on the offensive, writing angry papers defending the primacy of his work, and insulting other scientists who disagreed with or belittled him. "I might have omitted the word *stupid*," he wrote in his memoirs of one such assault, "but the criticism was quite sound and valuable."[54] "Please don't believe too much of Grassi & Cos.' work," he begged one colleague. "Their actual observations have been of the slenderest and . . . the rest is eked out by aid of my reports."[55] "The work on human malaria is only of secondary importance," he added, a "mere detail" of his work on birds.[56]

Ross's vitriol worked. In the end, it was he, not Grassi, who was awarded the Nobel Prize in 1902.[57] But much time was wasted in the ugliness. "I hate war and publicity of this sort," Manson complained.[58] And in the meantime, the pace of scientific research into the mosquito question—and with it, *Homo sapiens'* first big chance to tackle malaria transmission—ground nearly to a halt.

Public skepticism about the mosquito theory of malaria transmission reigned. The notion that *Anopheles* mosquitoes transmitted malaria, alone, didn't fully explain common experiences with the disease. Did all species of *Anopheles* carry malaria? And if so, why was it that malaria raged in places where precious few *Anopheles* could be found, and was scarce in places where *Anopheles* rose in giant black flocks? This phenomenon was particularly obvious in Europe, where the malaria vector *Anopheles maculipennis* appeared to have no relation whatsoever to the prevalence of malaria. European malariologists called the conundrum "anophelism without malaria." Further, was the mosquito the sole vector of malaria? If so, why was it that in places such as the Netherlands and Germany, fevers started in April and May, months before fat and sleepy *Anopheles* awoke from hibernation?[59]

Ross and Grassi's discovery shed little light on such questions. As a result, many thinkers held that if the mosquito did transmit malaria, there were other, still-undiscovered factors. According to this way of thinking, Ross and Grassi had found just one route among many, and perhaps not even the most important one at that. (Grassi called the elusive transmitter "Factor X."[60])

Countertheories abounded. A paper in the *Indian Medical Gazette* argued that "proper, filtered water supply" played a bigger role in malaria transmission than "the mere presence of *Anopheles*."[61]

The mosquito theory was just a fad, wrote another author, who claimed to be able to prove with statistical certainty that malaria was a waterborne complaint.[62] In fact, according to another, malaria actually was just a "disorder of degenerating white blood corpuscles."[63]

Ross and Manson's response to the missing pieces in their mosquito theory only inflamed the public's skepticism. Manson organized some poorly received experiments, which convinced few of anything. Meanwhile, Ross wildly proclaimed that he'd be able to extinguish malaria in every city in the tropics within two years.[64]

Ross's exaggerations provoked even more exaggerated dismissals among health officials and scientists, who pointed out the folly of attempting to effect mosquito genocides. "I doubt whether, even if the whole population of India were put to the work of filling up all the puddles during the rains, the results would justify the expense," maintained a British official at a malaria conference in India.[65] Indeed, the chairman of London's Society of Arts said to Ross, "Mosquitoes are everywhere. They surround us like the air we breathe."[66]

In fact, to control malaria transmission, it isn't necessary to slaughter every last mosquito or avoid every last mosquito bite. Only certain species of the *Anopheles* mosquito need to be tackled, and even then only in ways that make it difficult for them to transmit the malaria parasite. As Manson pointed out, getting bitten ten times a night in a place where one out of every one thousand *Anopheles* is infected results in three malaria infections a year. Reducing those bites to one a week would result in just one bout of malaria in *ten years*.[67]

In the polarized debate Ross sparked, however, such nuances were quickly lost. "The identification of mosquitoes has become so difficult that it is better to leave it alone," said the bacteriologist Robert Koch, "so long as there remains anything else to be done in this world."[68] Even in Italy there was little interest in controlling mosquitoes. Nobody seriously considered that all the Italian peasants sleeping in caves in the Campagna could be rid of every last mosquito bite. It was impractical, inconceivable.[69] According to the Indian

Medical Service's malaria expert Sydney Price James and the Dutch malariologist N. H. Swellengrebel, mosquito killing was "futile" and a "tyranny . . . over men's minds" that should be "thrown off," as they jointly reported for the League of Nations' Malaria Commission in 1927.[70]

Ross died in 1932 a bitter, ruined man. "I had lost money over the work, I had received practically nothing but skepticism or even abuse in return, and most of my results were credited to others," he wrote in his memoirs. "The word 'malaria'" he declared, made him "nearly as sick as the thing itself would have done."[71] To add insult to injury, the species of *Anopheles* named after him—*Anopheles rossi*—for years was believed not to transmit malaria.[72] When it was finally discovered that it did, it had by then come to be known as *Anopheles subpictus*, the name bestowed upon it by Ross's archrival Giovanni Grassi.[73]

In the decades following Ross's and Grassi's discovery, only those willing to flout the scientific conventional wisdom did anything much to minimize the reproduction or biting of malarial mosquitoes. The Italian prime minister Benito Mussolini did, during the 1930s, as a cornerstone of his Fascist revolution. His 549-million-lire scheme to finally drain the Pontine swamps took the lives of more than three thousand workers.[74] Industrialists with malaria-threatened rubber plantations in Malaysia did, discovering that disrupting malaria transmission sent other infectious diseases—diarrhea, dysentery, nephritis, abscesses, tuberculosis, convulsions —plummeting as well.[75]

So, too, did American malariologists. Insect destruction was generally quite popular in the rapidly industrializing United States of the late nineteenth and early twentieth centuries. American farmers, having turned the wilds into feeding troughs for insects shorn of their natural enemies, lived in fear of entomological invasion. Clouds of locusts descended upon the farms of the Mississippi Valley. Gypsy moths stripped every tree in New England of their leaves, and their decaying, crushed bodies sent a stench across the land.[76]

With insects strangling American economic development, the government had started to invest in entomological research. An early result, in 1901, was an influential handbook called *Mosquitoes: How They Live, How They Are Classified, and How They May Be Destroyed.*[77] Whether or not anti-mosquito campaigns alleviated malaria, they almost always helped improve property values, spur development, and after the First World War, showcase new chemicals, many of which had been refined for use as chemical weapons.

New Jersey state entomologist John B. Smith demolished acres of mosquito habitat in that state's lowlands, a feat that he claimed more than quadrupled property values. A similar campaign waged on Staten Island, in New York City, triggered a development boom.[78] In the early years of the twentieth century, the notorious oil baron John D. Rockefeller created the philanthropic behemoth the Rockefeller Foundation, whose International Health Division quickly gravitated toward the popular anti-mosquito projects. "I can't recollect when we have been able to remain on our front porches without fighting the blood sucker until this summer," one grateful local wrote to the foundation. "Thank you again."[79]

None of this budged the scientific consensus against mosquito killing in Europe. The Rockefeller Foundation malariologists made the rounds at international scientific meetings, touting their entomological victories, but with notoriously poor data collection and a chronic lack of controlled comparisons, Rockefeller scientists' optimistic reports left their European colleagues entirely unmoved. The reports didn't measure whether malaria had declined, nor the complicated role of the many confounding factors—including rainfall, temperature, and the movement of human populations—that may have played a role in any observed changes. In 1927, the League of Nations prepared a report dripping with skepticism about the Americans' claims.[80]

Rather than debate the issue and refine their science, the Americans, like Ross and Manson, closed ranks. Enraged American malariologists felt that their dismissive European colleagues couldn't be bothered doing anything about malaria.[81] The U.S. surgeon general pressured the League of Nations (which the United States had never joined) to bury the doubting report, claiming it suffered from "contradictory and often insufficient premises"[82]—and that was that. The international body never published it.[83]

Malariologists' dispute over the antimalarial utility of mosquito killing rested on differing conceptions of the nature of mosquitoes' carriage of malaria. Simply put, those who believed that all *Anopheles* mosquitoes carried malaria assumed that killing any *Anopheles* mosquitoes would help reduce malaria. Those who believed that something other than *Anopheles* mosquitoes carried malaria presumed that mosquito killing would do little to reduce malaria.

The truth was that neither side in the debate had it right: it wasn't that all *Anopheles* carried malaria, or that something other than *Anopheles* carried malaria, but rather that *only some Anopheles* carried malaria. Research into the biology and mysterious habits of *Anopheles* mosquitoes would have revealed this fact, but both the skeptics and the enthusiasts had deflated funding interest with their conflicting orthodoxies. The American entomologist L. O. Howard, for example, tried to launch a research project on the identification of different species of *Anopheles*. But "there were many more species of mosquitoes than I had supposed," with "infinite variations in habit," he remembered. And "we could not possibly produce such a work as we wished to bring out" on the limited three-year grant Howard had scraped together. Funds to conduct this kind of mosquito research were so thin that the only way Howard figured he could continue the work at all was by dipping into the deep pockets of a fellow biologist.[84]

And so clues, when they appeared, arrived from obscure, unheralded corners.

First, in 1921, the French entomologist Emile Roubaud speculated that malaria transmission might be linked to some hidden quirk inside the *Anopheles* mosquito. Say, for example, that malaria-carrying *Anopheles* in some localities had more teeth, he theorized. They'd be able to bite through animals' thick hides and therefore would deposit any malaria parasites within them into the dead-end host. Perhaps some other locality harbored mosquitoes with fewer teeth. Those insects would have no choice but to bite thin-skinned humans, and so could effectively carry malaria. The dental difference would explain why not all *Anopheles* species, such as the European *Anopheles maculipennis*, seemed to carry malaria.

Or say, as a retired public health inspector in Italy did in 1926, that the *Anopheles* mosquitoes that laid gray eggs carried malaria, while those that laid, say, dark eggs did not. The inspector, Domenico Falleroni, collected mosquito eggs as a hobby, and had noticed that individual female mosquitoes always laid eggs with the same markings. Finding the delicately designed eggs quite beautiful, Falleroni painstakingly described and categorized them, even naming two types *messeae* and *labranchiae*, after his friends from the health department Drs. Messea and Labranca.

Some scientists apparently conducted dental examinations of local mosquitoes and, finding nothing, dropped Roubaud's idea. Falleroni's ideas about the markings of eggs they dismissed as a "mere eccentricity of nature," as biologically insignificant as "spots on mongrel puppies," as the Rockefeller Foundation's Lewis Hackett put it.[85] It wasn't until the late 1930s that a collaboration between Italian scientists and malariologists from the Rockefeller Foundation revealed that what had been inexpertly termed *Anopheles maculipennis* was in fact five different species of *Anopheles* mosquito, visually indistinguishable except for the delicate markings of their eggs. Of the five, only Falleroni's *Anopheles messeae* and *Anopheles labranchiae* transmitted the scourge; the others were blameless.

• • •

This finding resolved the scientific impasse that had impeded widespread acceptance of the mosquito's role in malaria for some forty years, decoding the mystery of millions of cases of European malaria.

Even more than that, it ushered in a new paradigm for malariology. Scientists realized that deciphering the ecology of malaria transmission required much more than simply checking the *Anopheles* credentials of the local mosquitoes. In Europe, delicate mosquito eggs had to be gathered and studied. In the United States, the two tiny hairs that protrude from mosquito larvae's heads had to be examined—on harmless *Anopheles punctipennis*, those hairs are close together at the base; on the killer *Anopheles quadrimaculatus*, they're spread ever so slightly apart.[86]

As the morphological differences between *Anopheles* species became clear, so did their unique habits, and the specificity with which each would have to be stalked within its own ecological niche. To stanch transmission, local entomologists had to work with engineers, who had to work with health officers and clinicians. For, as Lewis Hackett put it, "the best method in one place may be the worst possible thing to do only forty miles away."[87]

Hackett summarized the new thinking in 1937:

> A mosquito, harmless in Java, is found to be the chief vector in the interior of Sumatra. A method of treatment unusually successful in India is almost without effect in Sardinia. The half-mile radius, sufficient for larval control in Malaya, has to be quintupled in the Mediterranean basin. A village in Spain, in which half the population is in bed with chills and fevers in August, turns out to be less infected than a village in Africa where virtually no one has to abandon work on account of malaria at any time.[88]

Malaria, he said, was "so moulded and altered by local conditions that it becomes a thousand different diseases and epidemiological puzzles."[89] Each would have to be unraveled on its own terms.

• • •

It took malariology four decades to grasp the futility of single-bullet solutions to malaria. And yet, the paradigm-shifting insight described by Hackett has mostly been lost.

In part that's because of the vagaries of malaria research funding. Local, ecologically driven malaria research is not particularly applicable to other areas of the economy, nor to other areas of the world. It must be funded locally for public health reasons alone, and political will or financial resources are lacking for it in most malaria-endemic regions. It's challenging enough to fund proven treatment and prevention, let alone in-depth investigations into the entomology, ecology, and epidemiology of local malaria transmission. Even in wealthy countries, support for malariology ebbs and flows. When Italian authorities believed they'd solved their domestic malaria problem in the 1920s, for example, they dismantled their malaria research infrastructure altogether. When the United States and other international public health authorities believed DDT and chloroquine would end malaria, they similarly stopped funding research.

It's also because, by the time Hackett's revelation emerged, much of the infrastructure for malaria science had largely been built, and it suited locally grounded, ecologically minded malariology as well as a shoe fit a hand. None of the malaria research centers established by Patrick Manson, authorities in the British Raj, or the Rockefeller Foundation were sited in malaria-endemic regions, where malariologists could study malaria up close and on the ground. With the financial support of colonial authorities, Manson helped found the London School of Hygiene and Tropical Medicine in the malaria-free city of London.[90] The Raj built malaria research institutes in India's cool hill towns, which Britishers found more comfortable but where malaria seldom arose.[91] The Rockefeller Foundation poured dollars and the expertise of its malariologists into public health research at American universities, such as Johns Hopkins and Harvard.[92] These centers form the backbone of global malaria research to this day.

As a result, the thrust of the most high-profile and well-funded malaria research is devised by specialists oceans away from wild malaria, intended for use everywhere regardless of local malaria ecologies, and sponsored by funders intent on bold gestures, not idiosyncratic tinkering. It's the very opposite of locally tailored.

Take the boom in malaria vaccine research.

There are now dozens of experimental malaria vaccines percolating in labs across the globe. Vaccine research is expensive, and progress has been limited. The vaccine at the most advanced stage of development, called Mosquirix, was first created by scientists at GlaxoSmithKline from a tiny piece of the falciparum parasite, specifically a subunit of a protein of the sporozoite. So-called subunit vaccines, while considered safer than vaccines made from whole pathogens, don't generally trigger particularly vigorous or long-lasting immune responses. Mosquirix is no exception. Clinical trial results released at the end of 2008 showed that Mosquirix reduced the incidence of infection by 65 percent and of clinical malaria by nearly 60 percent, but for only six months.[93]

Another malaria vaccine candidate, created from a whole sporozoite damaged by radiation, has triggered complete immunity, but only in nonimmune adults exposed to experimental malaria in the lab. That vaccine, under development by a new company called Sanaria, suffers from fierce manufacturing and distribution headaches. As of 2008, the sporozoites have to be raised in live mosquitoes, and vaccination would entail injecting people with up to ten thousand live and kicking *P. falciparum* sporozoites, an unpopular approach that could badly backfire.[94] Most of the other vaccines are aimed at blood-stage parasites. Such vaccines can reduce illness, but they won't prevent infection or interrupt transmission.

These results are not surprising. Effective vaccines such as those against yellow fever and smallpox are based on the fact that the immune system can naturally create perfect immunity to those pathogens if exposed at a low level first. That's why in places where yellow fever and smallpox have been endemic, many local people naturally acquire complete immunity. The immune response against malaria is neither as complete nor as long-lasting.[95]

Plus, malaria vaccines must face the challenge of all the multiple forms of the malaria parasite, each of which plays a different role in the business of malaria illness and transmission. A vaccine that helped the body fight off sporozoites could help prevent infection, but that same vaccine would not help the body battle merozoites, and so could not prevent illness, nor would it help fight gametocytes and thus prevent transmission (unless it provided extraordinary, 100 percent effective protection). Moreover, a vaccine that acted specifically against *P. falciparum* could not be expected to exert any action against *P. vivax*, or *P. malariae*, or *P. ovale*, not to mention other strains of *P. falciparum*.

That's why in the 1960s, experts at WHO and USAID decided against launching malaria vaccine research, and in 2007, a World Bank–sponsored report on malaria vaccines pronounced that "failure will continue to be the norm rather than the exception."[96]

And yet, malaria vaccine research is one of the most lushly funded and high-profile areas of malaria research today. Why? Because a vaccine is the ultimate single-bullet solution: one shot that would bestow upon its recipient complete lifetime protection against *Plasmodium*. A vaccine would dispense with complex ecological conditions and convoluted malaria epidemiology. Distribution wouldn't even require a health clinic. Thousands could be vaccinated in a matter of days, cheaply, via traveling makeshift camps. Since the late 1990s, the Bill and Melinda Gates Foundation, which promotes bold, technically difficult solutions, has devoted $150 million exclusively to the search for the malaria vaccine.

Or consider the high-tech genomics-based drug development research that Dyann Wirth and her colleagues conduct at the Harvard Malaria Initiative. Wirth studies the genetic diversity of *P. fal-*

ciparum, charting its vast variability, in order to pinpoint parasite genes and proteins under siege by the human immune system. "It is very basic, very fundamental" kind of work, Wirth says. If successful, the research will reveal potential weak spots inside the parasite.[97] Then molecular biologists may be able to synthesize a drug to attack it.

If Hackett's vision of malariology was as an interdisciplinary, holistic science, the Harvard Malaria Initiative's method of drug development is the very definition of reductionist. Our best antimalarials—quinine and artemisinin—are nothing like the synthetic, highly targeted chemicals that HMI's model will eventually create. They're diffuse-acting compounds created by plants and discovered by traditional healers, which is in part why quinine has yet to provoke much resistance, despite centuries of use. HMI's drugs will pursue the parasite with intense, surgical precision, inevitably exerting pressure on it to evolve resistance. There are cheaper, more proven methods of finding new malaria medicines—for example, by tramping the world collecting traditional medicines, or screening thousands of compounds for antimalarial activity. That's how we've found most other drugs and all other antimalarial drugs.

But despite the fact that HMI's drug development model is unproven and will render the kind of drugs most likely to trigger resistance in the parasite, it, too, is lavishly funded. Why? Because it uses cutting-edge, economy-building technology. In 2002, scientists decoded the genomes of *P. falciparum* and *A. gambiae*, and their results were splashed dramatically across the covers of top scientific journals *Nature* and *Science* simultaneously. Some of the most well-endowed institutions in the world turned their attention to malaria then, many for the first time ever.[98]

It's hard to predict how long the current funding wave for malaria research will last, or what may follow it. Perhaps the wave will gather strength and be followed by another one, and another—a flood spilling over the land. Or perhaps, when enthusiasm for genomics or bold gestures inevitably crashes, the wave will quietly recede.

After my visit to Harvard, I descend to the underground parking lot to find my beat-up Honda amid the Audis and BMWs. I pass through a sun-filled atrium, where students and faculty lunch on green salads and beautifully ripened fruits, their backpacks and satchels slung over the backs of their chairs. Perhaps their aspirational, high-tech research will render just the kind of scientific solutions the malarious masses need. If the history of malariology is any guide, the next scientific breakthrough in malaria could come from anywhere at all. Tomorrow's antimalarial superweapon may well be lurking in the lab notebooks and journal papers that peek out of these researchers' bags.

Before pulling open the door, I draw thick mittens over my hands. The winter air outside the humming building is cold indeed.[99]

8. THE DISAPPEARED: HOW MALARIA VANISHED FROM THE WEST

Life for a malarial mosquito isn't easy inside London's Houses of Parliament these days. It's drafty. Sharp, clammy winds waft through the imposing building, which is cold and dimly lit. In the autumn of 2006, a few unfortunate mosquitoes, plucked from their adopted home in a local laboratory—where they were coddled in a specially heated and humidified insectary—braved the punishing conditions in service of a small and mostly ignored exhibit about malaria. The exhibitors positioned the mosquitoes inside an eighteen-inch glass cube placed on a table ringed by a few bulletin boards in a vast and empty hall tucked in some out-of-the-way corner of the building. The organizers had no doubt hoped the shivering insects would provide an edge of drama to the exhibit—real, live killer mosquitoes!—diluting the dispiriting eighth-grade-science-project effect of the Formica table and thumbtacked posters. But London being London, the heat had been turned off in the Parliament over the weekend, and the exhibit's six-legged headliners froze to death.

Now it was up to a cadre of mosquito specialists, summoned from the nearby London School of Hygiene and Tropical Medicine, to replenish the cube. Inside their sultry insectary, they lured a few captive *Anopheles gambiae* into a Dixie cup, placed a piece of gauze over the top, and secured it with a rubber band. Professor Chris Curtis tucked the cup and a few other tools into his handbag and, with malariologist Jo Lines, two research fellows, and a technician in tow, boarded the London Tube bound for the Parliament.

One of the sponsors of the exhibit—though the very picture of modesty, the exhibit had been sponsored by a raft of multinational outfits, including GlaxoSmithKline, Novartis, Royal Dutch Shell, and UNICEF—would meet the mosquito-laden scientists at the entrance to the building. Curtis called the benefactor, simply, a "very rich lady." Whisked through security ("What is *that*?" a guard exclaimed with some alarm, encountering Curtis's two-foot-long mosquito-sucking tube), the group headed straight for the expiring insects. A pile of mosquito corpses lay on a piece of filter paper at the bottom of the forlorn cube. ("So malaria has been eradicated!" joked Lines. Pause. "In Parliament.")

The research technician got to work. She pulled out a little sachet and dispatched one of the research fellows to the men's room to soak it in water. Meanwhile, Curtis rolled up his sleeve and fished out the Dixie cup, which he turned upside down and placed snugly over his forearm, so the gauze pressed against his skin.

Passersby murmured in wonder as Curtis encouraged the mosquitoes to feast on his blood. The mosquitoes rooted around on Curtis's arm frenetically. The journey and the strange new environment had chilled and rattled them, the technician, Shahida Begum, told me. They might refuse the blood meal altogether. As finely attuned to their needs as Begum was, there wasn't much she could do. They might not like the taste of Curtis, she whispered.

In the insectary, Begum could keep captive *Anopheles* alive for around thirty days. In the stiff and proper environs of the Houses of Parliament, despite the ministrations of a crack team of world-class experts, they'd be lucky if they lasted the night.

• • •

How the mighty have fallen! Earlier generations of *Anopheles* in London did not shrink in forgotten corners. They ruled the city, unleashing outbreaks of malaria that held the city in fevered thrall. Once, the powerful men in the Houses of Parliament quaked in their boots at the thought of the mosquito's wrath.

Granted, that was around four hundred years ago. By then, malaria had long prospered in England. *P. vivax* first arrived in Britain after the Roman Empire declined and the Roman technology that had held malaria at bay fell into ruin. By the end of the first millennium, *P. vivax* had sunk its tentacles deep into the low-lying marshlands around the Thames estuary. It didn't leave until the end of the nineteenth century.

Kent is today known as the Garden of England for its lovely orchards, and Essex, among other things, as a popular suburban area convenient to London, but as long as malaria ruled these two counties, nobody wanted to live there. Those who did were called marsh dwellers, and regularly suffered the "marsh ague." Although *ague* literally means "acute," and at the time could refer to any fever or malaise, the marsh ague was a very particular disease: a "rigor and horror which is succeeded by heat and that afterwards by a sweat," as the seventeenth-century English physician Thomas Sydenham described it, leaving little question as to its malarial nature. The marsh dwellers were "very rarely without" it, wrote the eighteenth-century historian Edward Hasted, "and if they survive, are generally afflicted with them till summer, and often for several years."

Horrified visitors noted the swollen spleens of the local children—which they dubbed "ague cake"—and the sallow complexions of the adults who lived with them. In the marsh counties "it is not unusual to see," wrote Hasted, "a poor man, his wife, and whole family of five or six children, hovering over their fire in their hovel, shaking with an ague all at the same time." It was "the moory soil, the watry atmosphere," an anonymous poet wrote, "With damp, unhealthy moisture . . . [and] thick, stinking fogs, and noxious vapours," that were to blame. "Agues and coughs are epidemicall; / Hence every face presented to our view / Looks of a pallid or a sallow hue."

The highland women whom the marsh men often married and brought back to their malarial homes succumbed with terrible regularity. "When the young lasses . . . came out of their native air into the marshes among the fogs and damps, there they presently changed their complexion, got an ague or two, and seldom held it above half a year or a year at most," wrote the seventeenth-century English novelist Daniel Defoe. Defoe claimed to have met men in Kent and Essex who'd lost more than a dozen wives to the marsh ague.[1]

Mortality rates in the infected counties rivaled those of present-day sub-Saharan Africa.[2] Other microbes swooped down in periodic epidemics that could carry off as much as 20 percent of the population. Despite the region's rich agricultural soil and its proximity to the coast and the commerce of London, fewer people lived in the marsh counties than anywhere else in England.[3] And those who remained were perhaps not England's finest. Vicars fled the afflicted marsh parishes, and the resulting lack of religious instruction, coupled with the local habit of treating malaria with liberal volumes of alcohol and opium, probably had something to do with it.[4]

The marsh counties' malaria also posed a danger to the packed masses in London, just upstream on the Thames. Sulphurous salt marshes stretched right across from the Palace of Westminster, and now and again, during an especially warm or wet season, infected *Anopheles* from the marsh counties would colonize the crowded city.[5] Sir Walter Raleigh, captive in the Tower of London on the Thames in 1592, "prayed fervently to God that he might not be seized with an ague-fit on the scaffold, lest his enemies should proclaim that he had met his death, shivering with fear."[6] In 1661, malaria outbreaks so roiled London that Parliament House declared a day of fasting to "pray for more seasonable weather."[7]

• • •

By the nineteenth century, Britain had more effective ways to counter malaria than divine entreaty. By then they knew the disease was linked to stagnant waters and could be alleviated with improved drainage. But even at the height of its powers, the British Empire did little to disrupt malaria. Political will to tackle the scourge coalesced only to protect a select few.

In Britain's West African colonies, British colonists were the chosen. Although British scientists and officers posted in African colonies pressed London's Colonial Office for supplies to improve drainage, which they felt certain would alleviate the ubiquitous malaria, Colonial Secretary Joseph Chamberlain decreed that rather than beat back malaria for the public's benefit, the British policy would be to sequester the Europeans as far away as possible from the mosquito-ridden lowlands and the malarious native subjects who lived there. Chamberlain sent confidential letters to the governors of the colonies with instructions that all new buildings be located "away from native quarters." "On no pretext whatever should residences be built near any native quarter, or that native quarters be allowed to spring up in the midst of the new European quarter," his advisors warned. "Otherwise the inevitable result will be that malarial fever, contracted from the native children, will be as rife, if not more so, as it is in Lagos."[8]

And so, above Freetown, the Crown built a segregated European-only enclave with one hundred tons of English cement and prefabricated bungalows. Called Hill Station, it boasted brand-new houses, a rail line, playing fields, exclusive clubs, and fresh, clean water running through pipes. The authorities banned Africans from settling anywhere within a mile radius of the new town, or from even entering any of the gated compounds. All of the clubs and fields were surrounded by a 440-yard-wide band in which no native people (except the "*bona fide* servants of Europeans") were allowed. The benefits of this selective defense against malaria rippled through the culture. Hill Station residents avoided not just malaria but the indignity of living near "dirty, overcrowded" native huts. They enjoyed quiet nights, free of the buzz of the mosquito as well as the "drumming and other noises dear to the Native."[9]

During World War I, British soldiers were the chosen few. Like sleepwalkers snapping to wakefulness, some of the same medical officials who expressed apathy about taming malaria in Africa threw themselves into robust and effective antimalaria campaigns to protect the troops. They conducted surveys; they mapped malaria's geography. In all the places where British troops congregated, "innumerable ponds were filled in or drained," writes the medical historian Mark Harrison. "Where this was not possible, petroleum spraying of breeding pools was carried out." With the improved methods made possible by the discoveries of Ross and Grassi, they introduced fish to devour mosquito larvae, distributed mosquito nets, and dredged silt from riverbeds to wash away larvae.[10]

In contrast, in British-ruled India, the British knowingly worsened malaria. There, they built dams across scores of wild Indian riverways, creating thousands of miles of irrigation canals.[11] The irrigated farmlands were better for wheat, sugarcane, cotton, indigo, and opium—export crops the British authorities could tax—than for the locals' traditional sustenance crops. And the East India Company could charge locals five rupees for every acre of farmland watered by the new canals. Plus, as one commentator put it, the transformed Indian agriculture would better "preserve the people from starvation."[12]

But locals complained that bouts of malaria followed in the wake of the new canals. They knew the drawbacks to their traditional irrigation schemes, such as inundation canals, which were labor-intensive and only seasonally useful.[13] About a quarter of the harvest was lost every year due to variable rainfall, and rural Indians lived in fear of sporadic famines.[14] But Indian farmers famously conserved the waters they did collect. This was holy stuff, after all: the water goddess Ganga incarnate.[15] Plus, traditional methods minimized disruption to malaria ecologies. The fish that swam through the inundation canals fed on *Anopheles* larvae.

An 1845 report by a British officer, T. E. Dempster, found that the new irrigation canals disrupted natural drainage patterns, with the result that when it rained, sheets of water spread across the plains, forming *Anopheles*-friendly bogs and swamps. "All our previous knowledge and experience would lead us to suspect some mischief from irrigating canals in such a climate as that of India," Dempster opined, "especially, if not expressly constructed so as to preserve the drainage of the country." And yet, the canals had been built "without any view to preserving the drainage of the country," and so had "every where impeded or absolutely obstructed it." As a result, the people who lived in the newly irrigated villages suffered a tremendous malarial burden, their spleens swollen at twice the rate as those in unirrigated or traditionally irrigated villages, Dempster reported.[16]

Rural, conquered Indian subjects, however, were not the chosen few. The construction of irrigation projects continued unchanged. When confronted with the extent of Indian malaria, British medical officers who'd commanded effective anti-mosquito campaigns to protect British troops, such as the Indian Medical Service's Samuel Rickard Christophers, emanated "an impression of indecisiveness," according to Christophers's fellow malariologists H. E. Shortt and P.C.C. Garnham.[17] British authorities refused to conduct another survey on malaria and irrigation, condemning the idea as "foolish" and a "wild scheme." By 1894, fever took the lives of five million Indians, about a quarter of them thanks to malaria.[18] In 1908, a malaria epidemic in heavily irrigated Punjab killed three hundred thousand over the course of a few months.[19]

British exclusivity in malaria control did not escape the notice of outside commentators. A 1929 League of Nations commission called the British malaria policy in India "sanitary inaction."[20] The nurse Florence Nightingale called it a failure of compassion. "We do not care for the people of India," she bemoaned. "How else to account for the facts about to be given? Do we even care enough about their daily lives of lingering deaths from causes we could remove?"[21] Medical officers who lived below Hill Station complained that antimalarial segregation undermined the British imperial mission to educate and civilize the natives.[22]

British authorities were unapologetic. Antimalarial segregation may not be "a just method, for the native is neglected," a top British medical official allowed, but it was "not right to sacrifice the European lives . . . to such sentimental ideas."[23]

For the nineteenth- and early twentieth-century British, the difficulty and cost of antimalarial measures, and the fact that the neglect of socially powerless people was considered a normal part of life, justified selective use of malaria-control techniques.

The calculus was the same across the pond, among the populist Americans.

As in Britain, malaria had spread across what would become the United States before anyone knew any better, as European settlers and their African slaves unknowingly contaminated the New World with *P. vivax* and *P. falciparum* parasites. After the American Revolution, malaria spread westward, rooting deep into the Mississippi Valley, an expanse of nearly continuous mosquito habitat that lay between the Appalachian and Rocky mountains.

Malaria took 80 percent of settlers in Pike County, Illinois, in the 1820s, and eighty of six hundred Norwegian settlers in Wisconsin in 1841. It destroyed an 1830s effort to build a canal between the Great Lakes and the Mississippi River. Cruising down the Mississippi, northern passengers ogled the "sallow faced . . . pitiable looking objects" that emerged from mud huts, sod houses, and dugouts along the river's banks. Their bloated spleens distended their bellies by nearly a foot. It was the "swamp devil," one boat captain explained. "I'm feared you will see plenty of it if you stay long in these parts," he said to a passenger.[24] "Don't go to Michigan, that land of ills," a mid-1800s song advised, "the word means ague, fever and chills."[25]

The disease barrier malaria created was so fierce that many felt the West would never be settled. Nevertheless, it wasn't the sallow-cheeked denizens of the Mississippi Valley who were the beneficiaries of the first purposeful U.S. campaign against malaria, which began in 1903.

The fight was waged much farther south, on the isthmus of Panama, where the United States hoped to build the canal that had eluded the French, the Scots, and the Spanish before them. To do it, they'd need to secure a healthful population of workers in one of the most malaria-infested parts of the continent. Here, too, the antimalarial techniques they enacted—in pursuit of economic goals—were restricted to a chosen few.

The man chosen to lead the attack on Panama's disease was a military doctor named William Crawford Gorgas. He'd never been particularly interested in health or medicine. Born to an elite family in pre–Civil War Alabama, as a child Gorgas fled Virginia barefoot as it burned, and so decided early in life that he would be a fighting man.[26] Getting into West Point was his "burning ambition," according to his biographer, but somehow the tall, soft-spoken Gorgas, even with all his connections (his grandfather had been the governor of Alabama, his father presided over the University of the South), couldn't secure a place there. This was a terrible blow, but the determined Gorgas discerned a back door: the military needed doctors.[27]

His father had been "aghast" at Gorgas's plans to enter medicine, even if it was just a ruse to get into West Point. A Gorgas going to medical school was sort of like a Kennedy driving a taxicab. This was decades before the age of antibiotics catapulted doctoring into a high art, before regulatory authorities required doctors to go through medical training, before Johns Hopkins opened its first medical school and required that its graduates accomplish college degrees and four years of training. Doctors were considered quacks, and medicine a joke—not surprising, really, given that medical treatment for even the most terrifying scourges such as yellow fever, which regularly rampaged the South during Gorgas's childhood, consisted of whisky, brandy, and cigars.[28]

By the turn of the century, Gorgas had a medical degree and a post in the army's medical department.[29] He first distinguished himself in the fight against yellow fever in Havana, Cuba, where he worked under Major Walter Reed, whose experimentation had established mosquitoes as the disease's vector. Having survived yellow fever in Texas, Gorgas enjoyed complete immunity to the virus. He didn't quite believe Reed's proposition about the bugs—Gorgas's anti–yellow fever method to date had involved burning down whole camps—but like any good soldier, he would not let personal belief eclipse the requirements of duty. Gorgas considered it his responsibility to "take precautions in this direction," and so he divided Havana into manageable districts, surveyed every house and family, and, with military precision, methodically destroyed every place he could find—every puddle, every open bucket—where the alleged mosquito enemy might find succor. By 1902, yellow fever was gone from Havana.[30]

With the Havana victory under his belt, Gorgas was the obvious choice for the job of attacking Panama's mosquitoes, whose role in transmitting malaria was by then six-year-old news. But most of the people with authority over the canal had yet to be convinced that mosquitoes, rather than smelly miasmas, transmitted the disease. President Roosevelt appointed his most accomplished engineers, the men who built the Washington Monument and the Illinois Central Railroad, to "make the dirt fly" in Panama, and pointedly did not name a medical specialist to the commission overseeing the canal's construction. Had it not been for a reference from a friend of Roosevelt's, Gorgas's assignment might never have happened at all.[31]

As a result, besides the powers of his considerable Southern charm (nurses called him "Dr. Gorgeous"), Gorgas had little real authority in Panama. He and his colleague Joseph LePrince arrived in Colón with a staff of just seven. They found stagnant water festering over the dilapidated town, and frame houses resting on piles above lagoons of green slime. Mosquitoes flitted everywhere: up and down the line of the proposed canal; well into Portobelo; teeming from rainwater barrels, streams, ponds, and swamps. And while there was no yellow fever to be seen (there were few foreigners in Panama to harbor it), the pallid hues and grotesque deformities caused by malarial parasites and mosquito-transmitted filarial worms abounded in the locals.[32]

Attacking Panama's resilient mosquito vectors would be a much bigger job than the assault on the yellow fever mosquito, *Aedes aegypti*, in Havana had been. *Aedes aegypti* lay its eggs exclusively in artificial containers of water found around human dwellings, making the species a pitifully soft target. In Panama, malaria burrowed as deeply into the isthmus's ecology as the spooky mangroves, whose roots rendered miles of Panama's swamps and shorelines impenetrable. *Anopheles albimanus*, which transmitted most of Panama's malaria, lay its eggs in fresh or salty water, in sunny puddles or lagoons, and fed on the blood of man or animal (although it particularly fancied horses), either indoors or out. No single point of attack could deter it for long. And like any creature well embedded in an ecosystem, *Plasmodium* enjoyed other alliances: the indiscriminate biters *Anopheles albitarsis* and *Anopheles punctimacula*, the saltwater *Anopheles aquasalis*, and the jungle mosquito *Anopheles darlingii* all ferried it into new bloodstreams when *A. albimanus* could not.[33]

Gorgas and his team set up shop in one of the French-built hospitals still languishing on the isthmus. Mosquitoes hatched from the potted plants in the hospital's elaborate gardens and sailed into the wards through unscreened windows to feed on the patients and their caretakers. Their tiny bodies covered the walls and doors of the hospital, each panel speckled with more than fifty at a time. After the sun went down, the staff could tend to patients only if a colleague stood beside them furiously fanning the mosquitoes away, while the night-duty nurses wilted, mummified in citronella-soaked bandages. Locals avoided the malarious hospital as much as possible.[34]

Gorgas immediately cabled Washington with a list of supplies. He requested great volumes of insecticidal pyrethrum and sulphur for fumigating, two tons of newspaper to seal the buildings from mosquitoes, masses of wire screening, a slew of workers, including at least twenty inspectors and one hundred trained nurses, plus start-up funds for a research lab.[35]

At first it seemed to Gorgas's team that their list of supplies must have glued itself to some "invisible spot" in Washington. They waited ("with the patience of angels," as LePrince put it). Finally, responses from Washington trickled in. Why was Gorgas sending so many costly cables? the officials in D.C. asked. He should use the mails instead. Wire screening and a research lab were out of the question, they informed him. As were the giant quantities of materials and labor. He'd get a quarter of the required volume of pyrethrum and sulphur, along with eight of the requested twenty inspectors and forty of the requested one hundred nurses. If he needed more workers, he'd have to "get a few niggers," they told him. Washington certainly wouldn't be sending any newspaper, either. In that case, they'd misunderstood Gorgas's meaning; they'd thought the good doctor wanted reading material.[36]

Dressed down, Gorgas dolefully made his rounds throughout Panama, watching the mosquitoes hatch and getting an earful of scorn from the engineers as he did. With his paltry supplies, Gorgas's staff focused on oiling puddles to smother mosquito larvae and fumigating the few buildings they could—in other words, making dirty puddles dirtier and stinky buildings stinkier. The canal commissioners were not impressed. "It's silly to spend all this money just to kill a few mosquitoes," one said.[37] "A dollar spent on sanitation is like throwing it into the bay," another added.[38] "On the mosquito you are simply wild," another said. "All who agree with you are wild. Get the idea out of your head."[39]

Gorgas's assault on Panama's mosquitoes began in earnest only after yellow fever broke out. Three quarters of all the Americans on the isthmus, including the chief engineer, fled. "A white man's a fool to go there," one refugee said, "and a bigger fool to stay." The Canal Commission members who traveled back and forth from Panama to the United States took to bringing their ornate coffins in tow, just in case.[40]

Alarmed, Roosevelt showered Gorgas with resources to do something about the mosquitoes.[41] (The canal commissioners and William Howard Taft, Roosevelt's secretary of war, still were not convinced; Taft complained that Gorgas had "no executive ability at all.") Soon Gorgas commanded a staff of 4,100 workers, a budget of $250,000, and the entire U.S. supply of sulphur, pyrethrum, and kerosene oil.[42]

The chosen few for his campaign would be the canal officials and those workers who lived within a narrow swath around the canal, called the Canal Zone. There, and in the parts of Panama City where canal officials lived and worked, Gorgas oiled puddles, fumigated buildings, screened houses, and drained swamps. He ensured that window and door screens remained in good condition, and assigned workers to stalk and smash adult mosquitoes by hand. He hired workers to manufacture a special pink soft drink loaded with quinine and serve it in the workers' mess halls, and still others to run dispensaries distributing free quinine. He wielded the authority to ban the keeping of cattle (which left puddle-forming hoofprints in the mud), to punish anyone caught harboring unsealed containers (which could fill with larvae-attracting rainwater), and to line workers up for quinine doses. It was a blitzkrieg.[43]

Beyond those circumscribed areas, malaria was allowed to feast unhindered. Despite their proximity to the Canal Zone, parasitized locals were considered outside Gorgas's purview. A series of malaria surveys in the 1930s, for example, found that more than 60 percent of villagers in towns within the canal's province, such as Portobelo and Nombre de Diós, carried malaria parasites. In other provinces, malaria rates among children ranged between 30 and 50 percent. In Darién, up to 70 percent were infected. "In a large part of this country the incidence of disease is probably as high as it ever was," wrote American military physician James Stevens Simmons in 1939. The Canal Zone was an oasis in the desert, "practically surrounded" by a "great prevalence of malaria."[44]

The anti-mosquito campaign excluded many canal workers, too, in part because they were the dark-skinned descendants of African slaves and thus relegated to second-class status. With few European or American workers willing to toil on the canal for a dollar a day, five sevenths of the canal work force hailed from the Caribbean, with forty-five thousand workers from Barbados alone.[45] The Canal Commission's explicit policy was to assign workers who were "undoubtedly black" or mixed race to "silver roll" status, and a much smaller group of white workers to "gold roll" status. While gold roll status entitled workers to plump paychecks, well-screened and fumigated cottages, clubhouses, and hotels and churches run by canal authorities, the silver roll workers got crammed into barracks, thirty of them to a five-hundred-cubic-foot interior,[46] sent to segregated hospitals, and barred from white facilities.[47]

Many fled the cramped barracks to take their chances outside well-heeled Panama City and the Canal Zone, hacking into mosquito-ridden jungles and swamps.[48] Photographs of their rough-hewn shacks show how vulnerable they would have been to insects during the prime biting hours, regardless of how mosquito-free the environments of their working hours might have been.

Gorgas himself encouraged black workers to abandon the Canal Zone and take shelter in the jungle, where none of his anti-mosquito methods could possibly benefit them.[49] In part, he thought that this would protect them from pneumonia, to which they seemed especially vulnerable. He also held, like many Southerners of his time, that black people shared a special biology that inured them to malaria and yellow fever. "The negroes seemed to resist the dangers of infection," he noted in a 1907 speech.[50]

In fact, just the opposite may have been true, for the tens of thousands of workers from Barbados, for example, would have had no immunity at all. Malaria was unknown on the island, so much so that for centuries Barbados was considered an antimalarial sanatorium.[51] It was canal workers from Panama who infected the island with malaria, triggering epidemics such as one in 1927, which took nearly two hundred lives.[52]

Nobody recorded the black canal workers' mortality and morbidity from malaria and other causes. Many of them simply left the isthmus, with daily steamers full of fleeing Caribbean workers departing with *Plasmodium* parasites in lieu of paychecks.[53] Labor historians such as Michael Conniff estimate that one out of ten black canal workers perished of disease and disfigurement,[54] a death rate four times that of the white workers whom they outnumbered by three to one.[55]

Unlike the unapologetically exclusionary Brits, however, populist Americans could not bring themselves to acknowledge Gorgas's de facto antimalarial segregation. Commentators dismissed the very notion of sickness among black canal workers. A *New York Times* reporter, for example, derided black canal workers seeking medical care as "lazy Negroes who are tired of working on the canal" trying to "look crazy enough to pass the examination before the hospital surgeon."[56] Visiting journalists and dignitaries scrunched up their noses at the workers' ramshackle settlements. "These people are of no more use than mosquitoes and buzzards," a member of a congressional committee remarked. "They ought all to be exterminated together." Whatever sickness they may have suffered stemmed from their own bad habits, Roosevelt opined. "A resolute effort should be made to teach the Negro some of the principles of personal hygiene," the president remarked.[57] Gorgas claimed to have tried. But "on the negroes," he explained, "we have difficulty in impressing the necessity of cleanliness."[58]

Fans exaggerated the extent of Gorgas's limited and expensive gains against the parasite. There was "nothing to match it in the history of human achievement," said influential physician William Osler, of Gorgas's work in Panama. "The whole world owes him a debt," said the *London Daily Mail*.[59] The Liverpool School of Tropical Medicine awarded Gorgas the Mary Kingsley Medal. The American Medical Association elected him its president. The University of Alabama also invited him to be its president, as did the University of the South. In 1914, he was appointed surgeon general of the U.S. Army.[60] A century later, Gorgas is still remembered as the "man who cleansed the Panama Canal of malaria," as *The New Yorker*'s Malcolm Gladwell wrote in 2001.[61]

Gorgas himself knew the limits of his achievement. "We did not get rid of malaria on the Isthmus of Panama, as we did at Havana," he admitted to a reporter. Rather, he'd shown simply that "the white man could flourish in the tropics."[62] And even there, to be fair, Gorgas could not claim full credit. During the building of the canal, level roads were laid and water and sewer systems established throughout the zone, as required by the terms of the treaty between the United States and Panama. Mosquito-infested puddles and rainwater containers diminished accordingly.[63]

Acknowledged or not, the Americans' selective antimalarial approach continued for years after the canal opened in 1914, even in Gorgas's home state of Alabama. Malaria in the U.S. South spiked in the first decades of the twentieth century, in the wake of a booming hydropower industry, which dammed dozens of wild rivers across the region, creating vast artificial lakes. Newly developed high-tension transmission lines meant that the rivers' pent-up power could be carried over vast distances to urban markets. But in many places, the dams worsened malaria. As the rivers disappeared, so did the little forest mosquitoes that lived in their shady running waters, such as *Anopheles punctipennis*, which hardly ever carried malaria. The resulting lakes were often clogged with branches, logs, and other debris, creating an unsightly mess rife with food and shelter for *Anopheles quadrimaculatus*, the region's most efficient malaria vector. Malaria parasites from all over the country arrived in the bodies of workers hired to build the dams.[64]

Hydropower companies didn't bother clearing the land before they flooded it, in part because reservoir clearing was expensive. Removing all the trees, rock, and brush along the sides of a river such as northern Alabama's Coosa, for example, could run to $60,000.[65]

As a result, when Alabama Power Company closed the dam it had built over the Coosa in 1914, acres of benign *A. punctipennis* habitat turned into malevolent *A. quadrimaculatus* territory, and malaria followed soon afterward. The year before the dam closed, there'd been twenty-five cases of malaria in the area. After, at least six hundred residents fell ill.[66] The local school closed down, the teacher so sick with fever he couldn't rouse himself out of bed. Fields of mature cotton lay unpicked, as sick and frightened sharecroppers fled their homes. "A poor man don't stand no more chance than a June bug in January," remembered sharecropper Willie Bass.[67] Even at the construction camp built by the power company, where employees enjoyed screened houses, sewage services, and regular medical attention, nearly every family sickened with malaria.[68]

Local health officers knew as well as the residents that the impounded water played a role in the malaria outbreak. "The more I see of malaria conditions," a U.S. public health service officer wrote to his boss in 1913, "the more I think that 'p-o-n-d' spells 'malaria.'"[69] "The fever come when the water was backed up over the trees," remembered Bass. "They began rotting and that put the fever in the air."[70] Similar transformations had wreaked misery upon the New England residents near milldams in the late eighteenth century. And entomologists had by then recorded precisely how subtle changes in larval habitats could transform malaria transmission by attracting more effective malaria vectors.

County health officials pleaded for funding to enact Gorgas's antimalarial methods. Malaria "exacts the heaviest tribute from the energies of our people," wrote one health officer in a 1914 report. "Will not the same kind of antimalarial work as was done on the Canal Zone bear the same kind of fruit if done around every household and every farm in Alabama?"[71]

But the power company denied any connection between its dam and the malaria. After all, it could see that many of the residents lived on properties rife with mosquito hatcheries, from muddy hog and cow pens to old wells, and couldn't figure out how a few more, along the edges of the new lake, could hurt. Plus, the residents' antipathy toward the company, which they didn't think would provide much electricity to them in the end anyway, was well known. Even before the dam closed, the "dam business" had been something of a "cuss word" for area residents, the local paper noted.[72] Company officials dismissed their complaints as the work of "rip-off artists looking for another way to make money without having to work," as the historian Harvey H. Jackson says.[73]

Enraged, the Coosa's fevered residents filed more than seven hundred lawsuits against Alabama Power, seeking over $3 million in damages. One resident sued the company for "the biggest figure of money he could think of." Area merchants banked on their impending victory, extending credit to customers who'd filed suits.[74]

Landrift Hand's case against the power company was one of the first to be heard. A rough-and-tumble sixty-five-year-old farmer, Hand lived on the river a few miles east of Shelby, Alabama. His land had always been boggy and mosquito-infested,[75] but everything had changed for Hand when the dam closed. Water flooded his land, its still edges lined with green scum, and a different, more malevolent type of mosquito infested his home. By the time the court heard his case, Hand had been suffering from malaria for weeks. Ears ringing from quinine, he'd been hit by a train he hadn't heard coming, mangling his right arm.[76]

Gorgas himself took the stand as an expert witness against Hand, on behalf of the company. He testified that Hand's malaria couldn't possibly have come from mosquitoes roosting in the dammed lake. Hand's domicile, some fifteen hundred yards away from the lake's edge, was too far away, Gorgas said, and the mosquitoes wouldn't have been able to fly that far. Rather, they must have emerged from the small dirty pools and puddles that littered Hand's property, just as the company maintained.

In fact, the lake's edge was the only viable larval site for the malaria-carrying mosquitoes that most likely bit Hand. *Anopheles quadrimaculatus* specialize in clear water with marginal vegetation, not dirty puddles. Entomologists had established as much years earlier, in a 1903 entomology book with which Gorgas was likely familiar. (The Harvard malariologist Andrew Spielman had a copy of the text on his bookshelf, signed by Gorgas's colleague LePrince.)[77] But in his testimony, Gorgas didn't distinguish *Anopheles* by species, which he said he considered as relevant as "the various subdivisions of the dog family."[78]

Nor was it true, as Gorgas testified, that *Anopheles* mosquitoes never flew more than two hundred yards from their birthplace. Military officers of the time had reported finding *Anopheles* two miles away from the bodies of water from which it hatched. Today, most mosquito biologists consider no less than five miles a safe distance.[79]

Nevertheless, Gorgas's testimony carried the day. After all, he was a national hero, an esteemed Alabaman, and the surgeon general of the U.S. Army. The Supreme Court of Alabama ruled that the Coosa's frightened locals suffered from "imaginary fears" of "imaginary dangers," and were not "qualified to form or have an opinion" on the condition of their flooded lands. The court threw out Hand's case after just a half hour of deliberation, and dismissed the 699 others.[80]

In 1916 and 1917, Alabama experienced "the greatest outbreak of malaria the state has ever seen," according to the state's health officer, S. W. Welch. "Not a family within three miles of the impounded water on both sides escaped malaria."[81] The state board of health felt helpless. "The stage is set in Alabama for epidemics of every known disease," Welch wrote in his annual report, "and your Board is helpless to prevent the impending catastrophe because it has no money."[82] There wasn't much they could do. Alabama Power had been authorized to build its dams by an act of Congress, and its activities were sanctioned by the most famous doctor in the world.[83]

The selective antimalarial approach of nineteenth-century Britain and the early twentieth-century United States could have vanquished malaria. Aggressively stanching the disease for a select population in a circumscribed area allows the malaria-free to achieve important political or economic goals. Through this limited strategy, wars are won and engineering feats are accomplished, which brings greater security and prosperity, which in turn allows for more malaria control, and more prosperity, and so on. Some modern malariologists, such as Harvard's Spielman, think so. It's the trickle-down theory of malaria control.

We'll never know. For in both the United States and Britain, malaria receded on its own, thanks to demographic and agricultural changes that disrupted malarial ecologies.

In the United States, the first rumblings occurred in the 1830s, when Americans discovered the rich seams of coal underlying their vast continent. The industrial development that coal unleashed paved over acres of mosquito habitat. Hundreds of thousands of miles of railroad track laid across the country for new coal-fired steam engines allowed people to move themselves, their homes, and their goods away from the *Anopheles*-dense riversides and toward the rails. Coal-fired factories sited conveniently close to workers and markets rapidly eclipsed the quaint riverside facilities, and the *Anopheles*-laden millponds that powered them.[84]

In the Mississippi Valley, farmers agitated for drainage schemes to rid the Midwest of its sixty million acres of wetlands, those squishy barriers to agricultural expansion that slowed horse travel around Chicago to as little as twelve miles a day. The wetlands were "solely tenanted by every worthless specimen of amphibious, vegetable, and animal creation," an 1847 farm journal article complained. Eradicating them by drainage was "all that is necessary to secure millions of acres" for wholesome, productive agriculture.[85] First by burying U-shaped tiles underground and later with industrial-strength machines, farmers shipped the pesky water off their lands into ditches and then into rivers and streams. Soon the valley's wetlands "began to look just like the other farmland," writes the environmental historian Ann Vileisis. "In a matter of generations, farmers would even forget where the tiles were laid unless they kept their grandfather's drainage plans in a desk drawer." As the spongy land dried out, downstream flooding worsened, and waterfowl and shorebird populations plummeted.[86] Mosquito habitats steadily vanished. The dairy cows that dotted the region's farms diverted the few mosquitoes that remained, and malaria parasites died out in the ruminants' inhospitable bodies.

By the end of the 1930s, malaria was a thing of the past in the Upper Mississippi Valley.[87]

Economic uplift measures in the South sent malaria packing there, too.[88] In 1933, President Roosevelt signed the Agricultural Adjustment Act, which helped Southern farmers mechanize their farms. A stream of black sharecroppers, made redundant by the farm machines, abandoned their swampy cabins for the city. Malarial mosquitoes' access to their bodies declined accordingly.[89]

Roosevelt also created the Tennessee Valley Authority, which built nine dams across the Tennessee River. By then, federal officials had established antimalarial regulations for the hydropower industry, requiring the clearing of potential reservoirs, the modulation of the level of impounded waters to alternately strand and wash away mosquito larvae, and the diagnosis and treatment of malaria-infected workers. (The industry's vulnerability to the animus of the "species *Homo*," who are "willing to go to any length to harass and extract money," as one public health official noted, had inspired the regulations.)[90] The TVA's cheap, plentiful electricity enriched the region, allowing for better roads and housing, which further reduced the mosquito's habitat and protected humans from its bite.[91]

By the time Rockefeller Foundation malariologists started staging popular anti-mosquito projects, and the United States created the Malaria Control in War Areas program in 1942 (which would later become the Centers for Disease Control), the weaknesses of their antimalarial methods didn't matter anymore. Malaria had already nearly vanished.[92]

Agricultural, economic, and demographic shifts broke malaria in Britain, too.

There, the first seismic shock occurred in the early eighteenth century, when a former politician named Charles Townshend started proselytizing to English farmers about a new agricultural method, borrowed from Holland, of planting four crops—wheat, barley, clover, and turnips—in constant rotation. For centuries, England's agricultural production had remained stubbornly stagnant. At any given moment, about one fifth of the country's arable land lay fallow, slowly recharging for another round of planting.[93] Medieval British farmers ritually slaughtered all their livestock on November 11, during the festival of Martinmas, before the cold set in, for they didn't produce enough to feed the animals over the winter. With four-crop rotation, no land had to stay fallow, and there'd be enough turnips to nourish livestock during the cold season.[94]

In 1670, fewer than 5 percent of farms in Norfolk and Suffolk grew turnips; by 1710, more than half did. Townshend's idea took off in malarious Essex, thanks to "the pursuit having become fashionable," one commentator drily noted.[95] Liberated from the annual cull, livestock populations grew. Between 1640 and 1730 in the Thames Valley and the uplands of Oxfordshire, flocks of sheep more than doubled in size, and the proportion of cattle herds over five animals strong grew from around a third to nearly half.[96]

As with the dairy cows in the American Midwest, the greater availability of cow and hog flesh attracted the interest of malarial mosquitoes. When faced with the exposed arm of a human or the flank of a calf, England's *Anopheles maculipennis* bit the calf four out of five times. Hovering between home and stable in search of a meal, she flew toward the stable ninety-nine out of one hundred times. That bit of anopheline caprice cost *Plasmodium* dearly.[97] As cows and sheep sprouted across the English countryside, malaria transmission ground to a halt. (Years later, when malariologists realized what had happened, some took to calling for "a pig under every bed" as an effective substitute for a mosquito net.[98])

Between 1700 and 1850, England's population boomed from fewer than six million to more than sixteen million,[99] despite regular outbreaks of typhus, typhoid, and diptheria, waves of cholera, and widespread tuberculosis, all untreatable by the medicine of the day.[100] England logged its last case of indigenous malaria in the final years of the nineteenth century.[101]

These stories do not appear on colorful posters about malaria, such as those that hung inside Parliament House in 2006. It's important to project an aura of agency, of the potency of collective will against indignations like killer mosquitoes in a modern world. It helps raise money, and it makes us feel better, too, about the mastery of our tools, the depth of our commitment, the power of our technology.

The truth is, the world's most powerful nations knowingly sacrificed whole generations to malaria's appetites. The pathogen ran away well before anyone thought to chase it.

9. THE SPRAY-GUN WAR

For most of humankind's history with malaria, political indifference, scientific controversy, and tightfistedness have reigned. The post–World War II development of a potent new compound called dichlorodiphenyltrichloroethane—or simply DDT—changed all that.

DDT unloosed the leash that had held us back, turning on its head every calculation about malaria and our ability to challenge it. A smoldering new collective resolve emerged. Political and scientific leaders from around the globe decided, en masse, to abandon its partial solutions: to stop trying to diminish malaria's burden, give up attempting to slow its progress or soften the suffering of its victims. Like some long-tormented creature exploding into a violent howl, they declared a fight to the finish. They'd use DDT to wipe *Plasmodium* off the face of the earth.

The outburst didn't last for long. But it changed the landscape of malaria forever.

Malaria had been a particular problem on the tropical battlefields of World War II. "Never before has this great disease predator had such an unsurpassed opportunity," complained one military official, in a 1944 issue of *Science* magazine.[1] The previous year, more than twenty thousand British troops wearing military-issue shorts had to be hospitalized for malaria during the invasion of Sicily.[2] At Bataan, in New Guinea, and in Guadalcanal, malaria sickened tens of thousands of troops, grounding whole divisions, felling more soldiers than enemy combat.

The German army purposely triggered malaria epidemics, such as in Italy in 1944. Drainage pumps on the Roman marshes ordinarily pumped excess water out to sea, desiccating the land enough to make it malaria-free and thus habitable for thriving cities and towns. By stilling the pumps, the Germans could have flooded the region and effectively impeded the Allies' progress. But German malariologist Erich Martini had studied the habits of the local malaria vector, *Anopheles labranchiae*, in depth, and he knew that inundating the region with the Mediterranean's salty waters would allow *A. labranchiae*, which can thrive in brackish water, to flourish. And so rather than simply stopping the pumps, they reversed them, salinating some ninety-eight thousand acres. Then they confiscated local stockpiles of antimalarial drugs.[3] As the German soldiers departed, they left behind "clever sketches," *The New York Times* reported in 1944, "of the plague of mosquitoes that would follow the flooding of the farmlands."[4] More than 100,000 of the 245,000 locals sickened with malaria.[5]

Across the United States and Europe, scientists toiled furiously to arrest the wartime spread of disease, and to find new products to replace those made inaccessible by the war. Among the new synthetic chemicals they unleashed was a range of insect killers that included an amazingly resilient compound made of carbon, hydrogen, and chlorine, a recipe first developed by Swiss scientists at the Geigy Corporation. A sample of the stuff arrived at the U.S. Department of Agriculture's entomology research station in Orlando, Florida, in the early 1940s, for testing. "Nothing had been seen like this before," remembers one malariologist.[6]

DDT had many remarkable qualities. Its effect was long lasting and relatively specific, with a special malignancy for small, cold-blooded creatures.[7] In DDT's presence, neurons would start to fire spasmodically, oblivious to countersignals from the brain, like an engine running without a driver. Jitters led to convulsions, which, if the dose was high enough, ended in death.[8] It wouldn't dissolve in water, which meant that DDT powder, even if sprinkled on human skin or inhaled, had no discernible effect on people. It also meant that it could persist in the environment, exerting its poisonous effect, for months.[9]

Older insecticides, made from flowers and metals, were difficult to produce, short-acting, and often so toxic they had to be arduously applied in small quantities by hand, for fear they'd destroy everything, not just weeds and pests. Safer, odorless DDT, by contrast, could be synthesized in factories.[10]

The notion that DDT could be used to exterminate entire species of living things first arose in the mostly malaria-free United States, where fed-up farmers and gardeners continued to battle insect pests.

When the U.S. War Production Board announced that small quantities of DDT would be made available for civilian use in August 1945,[11] everyone from homemakers to farmers to government officials jostled for a piece of DDT's diabolical magic. Gardeners and farmers were "raiding the stores for every can that shows its top above the counter," writes the medical historian James Whorton. DDT sales skyrocketed from $10 million in 1944, purchased mainly by the military, to over $110 million by 1951, mostly to farmers.[12]

And they loved it. "Never in the history of entomology," enthused the USDA's Sievert Rohwer, "has a chemical been discovered that offers such promise to mankind for relief from his insect problems as DDT."[13] *The New York Times* lauded the "Army's insect powder" as a magical compound "deadly to insects" and "harmless to man." Others likened DDT to lifesaving penicillin.[14] The U.S. secretary of agriculture proclaimed his dream that DDT and other insecticides might be seeded inside clouds, so that the chemicals would shower down with the rain.[15]

Why not? Americans eager for a taste of wartime glory could pursue insects to extermination, just as the Allies did the Nazis and the Japanese. The DDT war on insects would be "our next world war," *Popular Mechanics* announced, "a long and bitter battle to crush the creeping, wriggling, flying, burrowing billions whose numbers and depredations baffle human comprehension."[16] After all, noted the chief of the Chemical Warfare Service, "the fundamental principles of poisoning Japanese, insects, rats, bacteria and cancer are essentially the same." (The Nazi propagandist Joseph Goebbels had used a similar rationale. "Since the flea is not a pleasant animal we are not obliged to keep it . . . our duty is rather to exterminate it," he said. "Likewise with the Jew.") And the news media and government agencies often conflated the killing technologies of DDT and the atomic bomb, dropped on Japan just five days after DDT's public launch. *Time* magazine pictured Hiroshima's mushroom cloud next to news about DDT's debut.[17] The CDC published the same mushroom cloud on the cover of one of its publications, too, specifically to illustrate DDT's awesome powers, calling it "the atomic bomb of the insect world."[18]

In popular films and plays such as the 1948 radio play *Leinengen versus the Ants* (made into a film starring Charlton Heston in 1954) and the 1954 film *Them!* filmmakers portrayed insects as mindless mass killers overrunning the countryside just as Americans feared that Communists, Nazis, and other totalitarian types might. Insects were "an evil force," a member of the House of Representatives said, one that made people dissatisfied. "I do not need to tell you," he added, "that dissatisfaction breeds communism."[19] Exterminating commie pinko insects with DDT, in other words, became downright patriotic.

Insects' critical role in decomposition and pollination forgotten, entomologist E. O. Essig proclaimed in 1944 that "insects are enemies of man."[20] Why tolerate them at all? DDT had ushered in an "auspicious time," said the entomologist Clay Lyle in a 1947 address to a professional entomologists' society, for "determined campaigns" for "complete extermination."[21] The makers of DDT agreed. So did government entomologists. "We have the tools," the USDA's M. L. Clarkson told a congressional subcommittee, "to bring this to a final conclusion."[22]

DDT similarly inspired Rockefeller Foundation malariologists to ratchet up their own battles against malarial mosquitoes. For years, Lewis Hackett and others had been promoting the utility of attacks on mosquitoes. So had the forbidding Rockefeller malariologist Fred Soper. But Soper didn't believe in merely controlling mosquito populations, depressing their numbers sufficiently for malaria to decline or even die out, as Hackett did. He felt that every mosquito could—and should—be exterminated altogether. He'd done just that, he claimed, in Egypt and Brazil in the 1930s, both of which had been invaded by *Anopheles gambiae* mosquitoes from Africa. Soper boasted that he'd "annihilated" and "completely eradicated" the foreign mosquito, using the agricultural insecticide Paris Green. He felt that DDT, which he considered an "almost perfect insecticide," could be used for even grander mosquito-eradication schemes.[23]

Before the war, however, Soper had had only limited reach into global malaria territory. Brazil's Fascist leader, Getúlio Vargas, had acquiesced to Soper's methods, but elsewhere, authorities tended to resist his bold interventions. For one thing, his claims were exaggerated: ecological shifts probably played a role in limiting *A. gambiae*'s spread in Brazil, and *A. gambiae* had returned to Egypt by 1950.[24] Also, Soper wasn't the type to win any popularity contests. "The trouble with Soper," noted one military official, was that "he is not only personally a stinker but he is just plain dumb."[25] Even Soper's friend and Rockefeller colleague Paul Russell had to admit that many people described Soper "in terms I prefer not to quote."[26] American universities refused to hire him because they considered him a Fascist.[27]

But then, in 1944, the Allied victors established a new international agency, endowed with billions of dollars, to oversee a massive relief effort in places ravaged by the war. The United Nations Relief and Rehabilitation Administration (UNRRA) had over $150 million to fund medical work alone. And to direct the effort, the agency tapped Soper's boss, the head of the Rockefeller Foundation's health division.[28]

With the new deep pockets and friendly leadership of the UNRRA, Soper had his chance to attempt extermination on a grand scale.[29] With less than $3 million and two years, he proposed, he would rid the entire island of Sardinia of every last specimen of the local malaria vector, *Anopheles labranchiae*.[30]

Local leaders thrilled at the prospects. "The future will open up a completely different life for the island's generations," proclaimed one commentator at the time. "The stables will be filled with herds . . . the soil will become more fertile and the humus will give the cultivator all the fruits he deserves."

The campaign started in 1946. Sixty-five thousand workers doused the wild, rocky island with more than 250 tons of DDT.[31]

Soper's campaign in Sardinia echoed across the globe, as a planet-wide war against insect life unfolded. Insects everywhere found themselves under a chemical siege. Spray crews armed with canisters of DDT drenched the walls of more than a million American homes in the U.S. South,[32] and an army of pilots hired by the USDA and state officials blanketed the countryside with DDT and related compounds to exterminate fire ants, gypsy moths, sawflies, elm dark beetles, and pest mosquitoes.[33] In Greece, Venezuela, Sri Lanka,

Italy, and elsewhere, government officials, fed up with tedious environmental management and epidemiological sleuthing to minimize malarial mosquitoes, launched DDT assaults upon *Anopheles*. Trucks wafting plumes of DDT rumbled through the streets of Panama's Canal Zone every night; local children frolicked in the chemical clouds.[34]

Insect populations everywhere crashed precipitously. Hungry insects expired in the fields unfed, and agricultural yields doubled worldwide between 1947 and 1979 (thanks, too, to synthetic fertilizers).[35] In Greece, flies and lice were nowhere to be found, and the olive crop grew by 25 percent.[36] As mosquito populations were decimated, so were the malaria parasites within them. In Sardinia, cases dropped from more than 75,000 in 1946 to just 9 in 1951; in Sri Lanka, from 3 million in 1946 to 7,300 in 1956. Malaria was gone entirely from the United States by 1951.[37]

The dramatic fall of their longtime plasmodial foe inspired a messianic zeal in malariologists such as Soper and his Rockefeller Foundation colleague Paul Russell. Cut loose from the foundation's health division (which formally closed in 1951), Soper himself for a time headed the regional public health authority, the Pan American Sanitary Bureau (later known as PAHO, or the Pan American Health Organization).[38] And he wrangled a spot for his old Brazilian colleague from the anti-*gambiae* campaign, Marcolino Candau, as the head of PAHO's newly formed counterpart, the World Health Organization.[39] Both he and Russell formally advised WHO,[40] which agreed to dispatch teams of experts to all corners of the malarial world to preach the DDT credo. By the early 1950s, WHO experts had launched more than seventy antimalarial DDT demonstration projects in Southeast Asia alone.[41]

One DDT convert summed up the new ethos among the international community of malariologists at a 1950 WHO meeting. Folding his hands before him, he intoned, "Let us spray."[42]

• • •

Even as the DDT hoopla reached a fever pitch, there were scientists who mused over how little, really, was known about the tenacious new compound. DDT certainly seemed benign, especially in contrast to the older, highly toxic pesticides it had replaced. It killed insects, yes, but slowly, and it had no immediate effect on humans or other creatures. It didn't even have an odor.[43]

And yet its amazing persistence in the environment gave some scientists pause. Military scientists knew, from early toxicity studies, that when ingested in large enough doses, DDT could destroy small mammals such as guinea pigs and rabbits. The army's James Simmons recalls finding the results of early DDT testing "somewhat alarming," and "rather startling."[44] DDT's effect "on other insect life, on pollination by insects, and on various biological balances is not well known," warned Rockefeller's Russell in 1945.[45] "There is a great deal that is yet to be learned about how to safely use DDT," a USDA entomologist added in *The New York Times* shortly after DDT's public launch.[46] Warned another government entomologist, darkly, "Biological deserts may be produced by heavy treatments of DDT."[47]

The military had chosen to move forward with DDT despite its alarming toxicity profile, since the limited campaigns they were considering would never expose larger creatures to dangerously high doses—and even if they did, such collateral damage mattered little during the course of a war.[48] By the time DDT came into much greater use, there was little interest in studying its impact on the environment and even less funding for such studies. With DDT fulfilling the long-held dreams of so many different sectors of society, the whole focus of entomology had shifted toward expanding and refining the use of pesticides, not fretting about their drawbacks or exploring possible alternatives.

Entomologists who studied non-pesticide-based methods—crop rotation, for example—were "ridiculed" by their peers as part of the "lunatic fringe," as the historian John Perkins notes.[49] And while the U.S. government required chemical companies to establish safe-use guidelines for products such as medicines, there were no such requirements for pesticides, which could be marketed to the public with virtually no prior testing.[50] Of course, regulatory laxity made sense in the years prior to 1945, when the entire U.S. production of all pesticides amounted to less than one hundred million pounds of materials that nobody had any interest in applying lavishly.[51] But by 1951, U.S. manufacturers produced one hundred million pounds of benign-seeming, easy-to-use DDT alone.[52]

And so nobody paused much to ponder the bizarre reports of insects strangely immune to DDT. In 1948, government entomologists in Orlando, Florida, had noticed that a sample of houseflies collected around local dairies seemed strangely tolerant of DDT, withstanding the toxin for much longer than other flies.[53] A WHO malariologist enjoying lunch at a country inn in Greece spied an *Anopheles* mosquito placidly reclining on a DDT-drenched wall.[54] Within a few years of DDT's arrival, reports of unnaturally tolerant flies and strangely oblivious mosquitoes popped up in Lebanon, Saudi Arabia, Egypt, El Salvador, and Greece.[55] But when scientists from the U.S. public health service discussed the reports at a meeting of the National Malaria Society in December 1948, all agreed: those bugs were outliers, freaks, mutants.[56]

As a weapon to fight malaria, DDT was achingly simple. A small team of semi-skilled workers could spread the stuff in even the most forgotten and remote areas, without benefit of doctors, nurses, scientists, clinics, or schools. It was orders of magnitude more effective than older methods, its potency undeniable to the most skeptical scientist. Even the most lauded antimalarial programs of the past couldn't compare to the fantastic results attained by DDT spraying.

But more than anything, DDT's efficacy against malarial mosquitoes seemed universal. Less than two decades earlier, Hackett had pronounced malaria a "thousand different diseases"; here was the best antimalaria method the world had ever seen, and amazingly it seemed to work *everywhere it was tried*, regardless of local variations in malaria epidemiology.

"This is the DDT era of malariology," Russell proclaimed to an audience at the London School of Hygiene and Tropical Medicine in 1953, in one of a series of lectures published the following year under the bold title *Man's Mastery of Malaria*. "For the first time," Russell noted, "it is economically feasible for nations, however underdeveloped and whatever the climate, to banish malaria completely from their borders."[57]

Regional public health agencies echoed Russell's call. The entire American continent should be freed of malaria for all time, the Pan American Sanitary Conference decreed in 1954.[58] So should the entire region of Asia, the Second Asian Malaria Conference agreed that same year.[59]

Russell wanted WHO to support eradicating malaria from the entire planet, too. But while WHO's expert advisors agreed that DDT could dramatically depress malaria transmission, they were far from unanimous on the notion that it could effect a complete, planet-wide genocide on *Plasmodium*. Banishing malaria requires more than simply intensifying control efforts, just as sterilizing a kitchen floor requires more than simply better mopping. One must get down on hands and knees and target every last speck of dust, no matter how seemingly insignificant, from every crack and crevice. Eradication required a nearly impossible level of perfection.

A whisper of suspicions streamed from the expert advisors' reports, issued in 1951 and 1954. Think of all the mud-walled huts across Asia, they said. Would they absorb sufficient quantities of DDT? Wouldn't some mosquitoes escape unscathed, parasites hatching in their guts? What about all the people who slept outside, they asked, or those whom mosquitoes bit out of doors? Untouched by DDT, they'd remain silent carriers and purveyors of the parasite. Very few health programs around the world could even *control* malaria effectively. How many, suffering "critical shortages" of personnel, would be able to muster the organizational finesse to ferret out every last remaining parasite?[60]

But WHO was a fledgling organization, and it didn't have the luxury of abiding by the cautious consensus of its scientific advisors. In 1955 Russell spoke at a meeting of the political body that governed WHO, the World Health Assembly, telling the lawyers, unionists, and other political appointees there that WHO would be left in the dust if it didn't get on board, fast. Whatever WHO decided to do, he announced, a campaign for worldwide malaria eradication was already under way."[61]

Russell exaggerated. A wide range of countries had started using DDT to control malaria, but only a select few had announced their ambition to extinguish it altogether, and few of those had actually embarked on a purposeful eradication campaign.[62] Nevertheless, the assembly took heed of the esteemed malariologist's word, and instructed WHO to take the initiative in Russell's de facto global surge toward eradication. "There is . . . no other logical choice," the assembly's director-general said.[63]

Experts estimated at the time that turning DDT projects into a truly global effort would cost at least $500 million. WHO established a special malaria-eradication account to start collecting funds to do the job. After a year, aside from a UNICEF donation of $10 million, the account held a grand total of $63,000, from Germany, Taiwan, and Brunei. PAHO, which had trailblazed the agencies' calls for eradication, allocated only $193,000.[64]

That left the bold plan just $489,000,000 short.

By 1956, Paul Russell had met with the International Development Advisory Board, a State Department committee comprised of movie producers, newspaper moguls, and business leaders, and convinced them that the eradication scheme was not only relatively straightforward and scientifically urgent, but politically expedient as well.

The IDAB had been tasked with figuring out how to counter the

Soviet Union's magnanimous foreign aid programs in developing countries. Containing the spread of communism—and its acolytes' disinterest in buying American goods—had become a central political and economic preoccupation in Washington, and the State Department feared that the Soviet programs were winning hearts and minds and spreading the communist credo.[65]

According to Russell, bankrolling WHO's global malaria-eradication program was not only an "excellent investment" (since the wiping out of malaria would lead to economic development, greater land cultivation, cheaper exports, stronger demand for American products, and ultimately the end of poverty), but also, politically speaking, an "outstanding opportunity." Particularly in regions key in the fight against communism—Southeast Asia and the eastern Mediterranean—the biannual arrival of door-to-door DDT spray teams would provide "concrete evidence of the interest of the U.S. in the well being of the populations concerned," effectively neutralizing the Soviets' nefarious charm offensive.[66]

The advisory board agreed, and issued its plea for funding for the eradication scheme in a 1956 report. Malaria's demise could be had in just a handful of years, the board wrote. Zap houses with DDT twice a year for four years, surveil the area for another four years to ensure that no remnant malaria lingered, and extinguish those cases that did, and malaria would be kaput. After that, "normal health department activities can be depended upon," the board reported, "to deal with occasional introduced cases just as they now remain on guard against smallpox, cholera, and other diseases."[67]

There were several glitches to the plan, as WHO had outlined. The IDAB acknowledged few. Sure, the deepest jungles of the Amazon and the remote mountain villages of Asia might not be fully accessible to spray teams, as WHO fretted, but these malarious pockets did not pose a "significant threat," the IDAB report said. "No doubt malaria can and will be eradicated in these areas in due time." Nor did its report make much of the problem of mud walls or the fact that many people didn't live in fixed homes sprayable by

DDT teams. There was a "sufficient range of alternative techniques" to deal with such eventualities, the IDAB noted blithely.[68]

WHO had been making noises, too, about the difficulty of prosecuting malaria's eradication in places such as tropical Africa, which lacked roads, trained personnel, and other necessary infrastructure. "The most we may be able to do" in tropical Africa, the WHO director-general bemoaned, "is to ensure that everyone who has an attack of malaria can go and pick up some tablets."[69] With tropical Africa figuring relatively little in cold war machinations, Russell and IDAB recommended simply dropping the continent from the campaign altogether. Africa carried such little political weight in those days that they didn't even bother nodding toward the irony of continuing to call the campaign a "global" one, even while excluding a whole continent. Not to mention the danger of leaving untouched a giant reservoir of some of the most virulent malaria parasites in the world.

Nor did the IDAB report mention the possibility that the DDT campaign could actually cause more malaria deaths. Nobody disputed that DDT would depress malaria to begin with. But freed from chronic exposure, local people would quickly shed their acquired immunities. And then if, for any reason—lack of funds, lack of roads, lack of popular support—malaria returned, they'd be especially vulnerable. More people could die. A less-than-perfect eradication campaign, in other words, could end up being much deadlier than no eradication campaign at all.

The one glitch the IDAB report did acknowledge was the growing hordes of mosquitoes that could withstand DDT and its cousin compounds. The outliers who'd once seemed so odd had started breaking out onto the front pages. "Mosquitoes Developing an Armor Against DDT," *The New York Times* headlined in 1952.[70] Three years later, the paper was warning: DDT "in danger of losing the war" against malarial mosquitoes.[71] DDT-impervious mosquitoes were even turning up in places where mosquitoes had never been purposely targeted. In El Salvador, Greece, Lebanon, Iran, Saudi

Arabia, Indonesia, and Nigeria, DDT-spiked runoff from treated cotton and rice fields formed puddles in which mosquito larvae squirmed, emerging out of the water fully formed and entirely unmoved by the toxin.[72]

A few DDT-resistant mosquitoes here and there didn't make much difference in the campaign. DDT's neurotoxic effect depended upon its ability to bind to insects' nerve cells, so an insect with even very subtly altered nerve cells—a single amino acid out of whack, for example—could avoid DDT's worst effects. Even in places that had never seen DDT, there were likely a few such mutant individuals flitting around.[73] This was tolerable, though, for ending malaria transmission didn't require that every last mosquito perish. Indeed, on Sardinia, Soper's $11 million DDT blitz hadn't exterminated the entire mosquito population: it had simply depressed it long enough for malaria transmission to become impossible.[74]

But if DDT-resistant mosquitoes came to dominate a population of mosquitoes before malaria was fully exterminated, eradicating malaria with DDT would indeed become impossible. This had already happened in a little-known test project run by malariologists in Panama. They'd been spraying riverside villages with DDT and supplying antimalarial drugs to disrupt malaria, but after they logged an initial decline in malaria, cases started to climb again, as DDT-resistant mosquitoes took up the malarial cause after their susceptible cohorts died out.[75]

What this meant is that the eradication campaign would have to kill sufficient numbers of mosquitoes, and quickly enough—in less than six years, Russell figured, *before* large numbers of the mosquitoes became inured to the toxin—to make malaria transmission untenable. The DDT had to come on fast and heavy. "If countries, due to lack of funds, have to proceed slowly, resistance is almost certain to appear and eradication will become economically impossible," the IDAB wrote. "TIME IS OF THE ESSENCE" (emphasis in original).[76]

• • •

The political appeal of the campaign proved irresistible. One by one, the State Department, Congress, and President Eisenhower got on board. In 1958, in legislation introduced by then-senators John F. Kennedy and Hubert Humphrey, the U.S. Congress allocated $100 million for a five-year worldwide malaria-eradication program (and, in the same act, extended the Marshall Plan).[77] In his State of the Union address, President Eisenhower touted the program as a "great work of humanity" that would put the Soviets to shame.[78] It was "a Christmas gift directly to more than a billion people," *The New York Times* enthused.[79]

Multimillion-dollar checks went out to WHO, PAHO, and those countries willing to convert their malaria-control programs into eradication programs.[80] The developing world's biggest campaign was launched in India, with $38 million from the United States and $50 million from the national government.[81] Ninety-two other malarious countries devoted themselves to the eradication cause, with several allocating as much as 35 percent of their tiny public health budgets to sending teams of workers into the field, canisters of DDT tied onto their backs.[82] Between 1957 and 1963, the United States would spend $490 million on the campaign, its single largest contributor.[83]

For the first few years, the global malaria-eradication program was everything Paul Russell and IDAB said it would be. By 1960, eleven countries had banished malaria from their borders, and a dozen or more had sent malaria rates plummeting.[84] Cases across Central America and in demonstration villages in Papua New Guinea fell to negligible levels.[85] In India, an annual caseload of seventy-five million dropped to fewer than one hundred thousand.[86] In Borneo, the parasite rate fell from 35.6 percent to less than 2.0 percent.[87]

Newly malaria-free populations were as beasts of burden relieved of their loads. Life expectancy in Sri Lanka increased from forty-three to fifty-seven years, and the prime minister envisioned repopulating the uncultivable malarious parts of the island. Life expectancy in Sardinia increased from 30 percent below national rates in Italy to 3 percent above.[88] Rice cultivation increased tenfold in Greece, Morocco, and Indonesia. In Cambodia, land values doubled.[89]

According to the experts, it was only a matter of time before the disease would be no more and these gains secured for all time. "If such a degree of control can be obtained" as quickly as it had been in so many places, WHO reported, "complete eradication can be expected in the near future."[90] Showered with honors as the Man Who Ended Malaria, Paul Russell watched the proceedings from a new position at Harvard University, confident in an imminent victory.[91]

Across Europe and North America, tropical medicine departments closed their doors. What was the point of furrowing brows over a soon-to-be-extinct disease? The study of how insects transmit malaria and other diseases became "a dead field," said a Johns Hopkins malariologist. "DDT is killing it."[92] Malariologists drifted into other, more well-funded fields, and new young scientists did the same.[93] The nuanced, multidisciplinary field, enlisting the insights of engineers, entomologists, ecologists, clinicians, and anthropologists, had devolved into a single brute question: how to coat interior walls with two grams of chemical per square foot.[94]

Even before eradicating malaria, the joke went, the campaign eradicated malariologists.

The spraying continued.

According to the IDAB plan, every single domicile in the program had to be sprayed with DDT twice a year for at least four years. Across the globe, teams of workers appeared in villages and towns toting large canisters of strange-smelling chemicals. They demanded that residents leave their homes, and take all their food and eating utensils with them. They removed all the pictures from the walls and moved the furniture around, leaving the walls covered with a powdery residue. It smelled like chlorine.

Locals found that the DDT didn't just repel and poison *Anopheles* mosquitoes and the pesky houseflies and bedbugs. It killed their chickens, too, and, in Malaysia, the flies that parasitized the caterpillars that fed on thatched roofs. Not long after DDT spray teams left, the village roofs collapsed.[95] In Borneo, the DDT killed the cockroaches, which killed the cats that ate the cockroaches. The cat-free villages were left with a rampaging rat population, and their crop-destroying, disease-carrying ways.[96]

For all this and more, suspicion reigned. In North Vietnam, local revolutionaries suspected that the spray teams were somehow collecting military information with their strange equipment. Villagers feared that evil spirits would enter their huts through the small holes campaign staff had made to install mosquito-catching traps.[97] Buddhist leaders in Cambodia and earlier nonviolent leaders such as Mohandas Gandhi in India objected that the campaign offended their religious sensibilities.[98]

The benefits of malaria eradication would likely have outweighed many of these admitted downsides, but there'd never been any attempt to enlist the support of the malarious masses for the DDT campaign. From the beginning, the program had been a top-down affair, hatched, funded, and overseen by experts in distant lands. The residents whose homes would be doused with DDT had barely been asked for permission.

Even after suspicions and objections arose, campaign leaders did not call for a massive public education push or for greater local participation in the conduct of the campaign. Instead, they practiced damage control. In Borneo, WHO opened special centers to collect donations of new, healthy cats. In the most remote areas, it arranged for the Royal Air Force to air-drop, along with vegetable seeds and "4 cartons of stout for a recuperating chieftain," twenty cats ensconced in specially devised cat baskets.[99] ("Many thanks to R.A.F.," wrote one local, "also to cat donors and cat basket makers . . . very accurate dropping.")[100] In Vietnam, they organized soldiers to accompany the spray teams for protection.[101] To Gandhi and the Buddhists, they offered the following bit of convoluted logic: Nobody was forcing DDT upon the mosquito. "If she chose to break into his home to drink his blood," as one campaign malariologist said to Gandhi, "and died in the course of her trespass, that was her doing, not his."[102]

None of this proved particularly persuasive. "Operation Cat-Drop," as one bemused local called it, earned the lasting mockery of environmentalists everywhere (thanks in part to a fictionalized version of the story by the novelist T. C. Boyle).[103] Far from being reassured by the presence of troops, Vietnamese villagers groused that the DDT spraying was being forced upon them.

Before long, spray teams around the globe found themselves in villages where no one was home and every door locked. In India, fewer than one in nine spray teams adequately sprayed their assigned areas. In a village of sixty-three houses, ten doors would be locked, thirty-five residents would refuse access, and one house would be forgotten by the spray team. Of them all, only seventeen houses would be sprayed.[104]

Growing recalcitrance of the public took its toll on the morale of the spray teams themselves. Some started taking bribes from those who wanted their homes spared. Others grew weary of carrying their tanks around all day, and so doubled their morning spraying so they could rest in the afternoon. Still others sold their DDT on the black market, spraying homes with a worthless dilution instead.[105]

As the spraying stumbled, so did the business of surveillance. Surveillance was both a much bigger job than spraying, requiring more resources and infrastructure, and much more hidden from public view. Teams of workers had to go door to door, collecting blood samples and actively hunting for fever cases. Each slide of blood had to be meticulously examined for parasites, and if any were found, workers had to go back out and find and treat the host. Mosquitoes had to be collected and their susceptibility to DDT tracked. Few countries had the skilled workers, clinics, or support systems required.

Surveillance workers skipped remote and hard-to-reach villages and took extra blood samples from more accessible residents instead. Back in the labs, there'd be two- and three-month backlogs of unexamined blood slides towering over microscopists.[106]

Meanwhile, the *Anopheles* mosquito, having escaped the attention of antimalarial spray teams, continued to be exposed to low levels of the insecticide on DDT-doused crops, and its populations were growing increasingly tolerant of the toxin.

DDT use in agriculture, especially in developing countries, soared. International development experts pushed for more DDT coverage on the rice and cotton fields of the developing world, convinced that this would unleash the massive harvests required to end hunger and poverty. In 1952, American chemical companies sold twenty-five million pounds of DDT overseas. Over the following decades, its exportation more than tripled.[107]

Mosquitoes alit on the DDT-dusted vegetation, and in the DDT-contaminated streams and puddles, they laid their eggs. And what didn't kill them only made them stronger.

Solving these myriad problems required innovative, locally sensitive solutions, but with malariology all but dead, WHO was forced to dispense increasingly equivocal advice. Remembers the organization's José Nájera, "A solution was sought in oversimplification and standardization."[108]

WHO soft-pedaled the spread of insecticide-resistant mosquitoes, urging the spray teams to continue. In 1962, it claimed that resistant mosquitoes "in no case" put the prospect of eradication "in jeopardy."[109] (The Royal Society of Tropical Medicine and Hygiene reached the opposite conclusion that same year, claiming that resistant mosquitoes were "seriously interfering with progress" in the campaign.)[110] In 1970, WHO allowed that resistant mosquitoes challenged the outcome of eradication in a few countries, but called the problem "more of an inconvenience than a major obstacle."[111] In 1973, WHO advised countries to continue spraying—but to use more expensive alternatives instead.[112]

And the organization rubber-stamped mass medication programs. In Brazil, antimalaria leaders confiscated supplies of table salt from all commercial establishments and homes, replacing it with salt loaded with fifty milligrams of chloroquine per gram.[113] With WHO's blessing, leaders in Angola, Cambodia, French Guiana, Ghana, Guyana, Indonesia, Iran, Irian Jaya, the Philippines, Sarawak, Suriname, and Tanzania followed suit. Millions of people around the globe, whether infected or not, regularly drugged themselves with their daily bread.[114]

In most places, it took about six months of mass medication to trigger the emergence of drug-resistant parasites, sending malaria rates back up to their pre-medicated-salt levels.[115]

The United States's five-year funding commitment for the malaria eradication campaign came to a close in 1963.

Globally, annual malaria cases had fallen from 350 million to 100 million, a historic low, but everywhere one looked, poverty and malnourishment and instability still reigned.[116] Paul Russell and IDAB had claimed that malaria's demise would lead to greater prosperity and more land in cultivation, but when WHO hired an economist to describe just how, according to the historian Randall Packard, "no one was able to provide the data he needed."[117]

What Russell and IDAB hadn't figured on was the countereffect of rising numbers of surviving people. In Western countries, where malaria receded with the onset of industrial development, declining death rates were matched by declining birth rates. But when malaria was surgically excised from countries sprayed with DDT and other chemicals—leaving intact the unelectrified shacks and the landless peasants who lived in them—the result was quite different.

In Sri Lanka, for example, the population grew by over 3 percent a year between 1921 and 1975; the University of Michigan public health expert Peter Newman attributed as much as 60 percent of this growth to falling death rates thanks to the malaria-eradication effort. But birth rates did not decline, and the growing population, demanding food, medicine, and education, soon outstripped the modest gains in economic growth that malaria's decline had unleashed. As scholars pointed out, death rates had modernized, but birth rates remained ancient.

On Sardinia, the lives of the locals lengthened, but agricultural productivity and economic production continued a steady decline that had begun in the 1940s. Wealthy outsiders rented summer residences on the newly malaria-free island, but tourism didn't enrich the local Sards, many of whom fled the island in search of temporary jobs elsewhere.[118] "Their lot in life had improved little," writes the historian John Farley, "and only noncritical tourists wearing blinders could call what had happened progress . . . Tourists have certainly replaced mosquito vectors in Sardinia but the indigenous population remain second-class citizens."[119] The medical anthropologist Peter Brown calls the transformation of Sardinia "modernization without development."[120]

At the same time, the political calculus that set the funding stream aflow had shifted. In the United States, enthusiasm for bold chemical attacks on pestilence started to give way to fears of poisoning and overpopulation. By the 1960s, public health experts had reached a new consensus, that the most serious problem facing humanity was not excess death from disease, but just the opposite: overpopulation.[121] Under the new way of thinking, less malaria didn't mean more people to produce more food; it meant more people to eat food and use up scarce resources. Saving people from sickness just condemned them to death by starvation.[122] Wasn't malaria really a "blessing in disguise," the naturalist William Vogt posited, since "the malaria belt is not suited for agriculture, and the disease has helped to keep man from destroying it"?[123] Indeed, the United Nations' Food and Agriculture Organization's first world food survey, published in 1946, had laid blame for the malnourishment of over half the world's population on the decline in mortality from infectious disease.[124] As public apprehension grew, malaria eradicationists found themselves under attack for their shortsightedness.[125] Critics attacked Russell as a "dangerous doctor" whose ideas were "creating problems faster than they are solving them," as Russell put it.[126]

Public enthusiasm for DDT had soured, too. The first off notes sounded a few years after DDT's public launch, when the USDA admitted that the nation's milk had been tainted with the toxin. It turned out that many creatures that had at first seemed impervious to DDT had actually absorbed and stored minuscule amounts of it in their fat tissues. So long as fat tissue keeps DDT in stasis, it's a safe enough locale for the compound. But with a half-life of eight years, fat-ensconced DDT lasted long enough to persist throughout the food chain, and each creature that ate another received a full complement of its prey's lifetime stores of DDT. Some creatures high on the food chain accrued dangerous concentrations of the stuff in their bodies.[127] Robins, for example, had been eliminated entirely from a 185-acre plot at Michigan State University, thanks to DDT spraying against elm dark beetles and mosquitoes. The earthworms fed on the fallen leaves of the sprayed trees, and when robins ate the worms the following spring, they accumulated enough DDT to kill them, or to stymie reproduction for two years in those few who survived.[128] Cows that fed on DDT-dusted crops stored the DDT in their fat tissue, and secreted it into the milk that scores of American children poured over their morning bowls of Cheerios.[129]

By 1955, the nation had been coated with so much DDT that the typical American diet delivered 184 micrograms of DDT—.0002 of a lethal dose—every day.[130] That probably would have been unsettling enough. But that wasn't all. As the cold war had heated up, so, too, did American and Soviet testing of nuclear weapons, releasing invisible clouds of radioactive material into the stratosphere. By 1956, *Newsweek* magazine fretted, there was sufficient strontium 90 up there "to doom countless of the world's children to inescapable and incurable cancer." In 1959, *Consumer Reports* reported on a major study of strontium 90 concentrations in milk, and the film *On the Beach* terrorized mass audiences with its dark vision of a post–nuclear holocaust world.[131]

Fears of DDT raining down upon the nation emulsified with larger fears of mass poisoning by secret, invisible toxins. Nobody knew if bioaccumulated DDT actually posed any threat to human health,[132] but noting the dead birds rotting on their lawns, many Americans couldn't help but wonder: Would humans be next? After the publication of the biologist Rachel Carson's potent 1962 book, *Silent Spring*, which pointed out the folly of using widespread pesticides without a solid understanding of their health and environmental impacts, then-president Kennedy convened a committee that recommended, over the aggrieved howls of the chemicals companies, that the government phase out the use of their iconic DDT and other similar compounds.[133]

The United States's five-year allocation for the global DDT blitz against malaria ran dry a few months later. Nobody asked for any more.

With the abrupt end of the U.S. contribution, funds to the WHO special account for the malaria-eradication program "stopped cold," writes the historian James Webb.[134] USAID formally withdrew from the program; UNICEF halved its malaria staff.[135] For many countries, burdened by the expense and trouble of the campaign, this was just the excuse they needed. Soon some national governments were spending more on garbage collection.[136]

Malaria resurged.

From a low of 18 cases in 1963, malaria swept over Sri Lanka, sickening more that 500,000 in 1969.[137] Over roughly the same period the caseload in India zoomed from 50,000 to more than 1 million[138]; in Central America, from 70,000 to nearly 120,000[139]; in Afghanistan, from 2,300 to 20,000.[140] A year after Europe was declared free of malaria in 1975, a two-year malaria epidemic roiled Turkey.[141]

At their 1969 meeting, the World Health Assembly directed WHO to abandon the eradication effort.[142] The "dramatic recrudescence" of malaria "will not be possible to stop" without new national commitments, nowhere to be seen.[143] WHO dispatched teams to visit malaria programs around the world, advising them to switch from trying to eradicate malaria to learning to live with it.[144] In 1974, PAHO jumped ship, too.[145]

Paul Russell and Fred Soper were devastated. Russell avoided the subject altogether, remembers Andrew Spielman, "withdrawing from contact with students and faculty" at Harvard. Soper pretended it hadn't happened. When he wrote his memoirs, he made "no significant mention" whatsoever of the campaign he'd helped inspire.[146]

Between 1957 and 1967, the war against *Plasmodium* cost $1.4 billion or about $9 billion in 2009 dollars.[147] [148] To this day, malariologists and historians disagree over whether it was worth it. The optimists point out that, in just over a dozen years, malaria had been lifted from the shoulders of 32 percent of the human population, which is no small thing. And the first baby steps toward a public health infrastructure had been taken in some of the most remote corners of the planet. The maps that malaria-eradication teams created, for example, had proven crucial to the success of WHO's smallpox-eradication campaign, doubtlessly a high point for global public health.[149]

Where some see incremental progress, others see what WHO's Tibor Lepes described as "one of the greatest mistakes ever made in public health."[150] Malaria had been eradicated from just eighteen countries in the world, all of them either prosperous, socialist, or island nations.[151] That left some two billion souls still burdened with malaria.

And the malaria that stalked them was in almost every way more vicious and harder to control than it had been before the eradication effort.[152] Our best and cheapest weapons—chloroquine and DDT—had been rendered toothless. By the early 1960s, *Plasmodium falciparum* parasites resistant to chloroquine distributed in table salt had emerged in Colombia, Brazil, Venezuela, and Thailand.[153] Chloroquine-resistant *P. falciparum* arrived in Kenya and Tanzania by 1978, and spread throughout the continent within a decade.[154] Around the globe, thirty-eight species of *Anopheles* mosquitoes had developed resistance to either DDT or its cousin compound dieldrin.[155]

Worse, the shallow reservoir of public attention and political will to fight malaria had been spent. In 1979, WHO announced its triumphal success in wiping smallpox off the face of the earth,[156] and reoriented the organization away from attacking specific diseases and toward a commitment to providing basic health care to the masses instead.[157] The worldwide malaria-eradication campaign faded away to a soon-forgotten footnote. Not even WHO, which stopped using precise measures of malaria, even bothered seriously tracking the disease anymore.[158]

Some experts blamed the failure of the campaign on pesticide-heavy agriculture, which sped up insecticide resistance. Two Columbia University public health experts, for example, called malaria's post-campaign resurgence in India a "social cost" of growing high-yield crops.[159] Others blamed WHO for its lack of sensible leadership, particularly in the years after U.S. funding dried up. "All logic went out the window," complains Spielman.[160] Thanks to the skyrocketing price of oil, upon which its production depended, DDT's price had spiked near the end.[161] What if it hadn't? Russell blamed people in general. "Resistant strains of *Homo sapiens*" were the problem, he wrote, "impatient bureaucrats" and "deans of schools of public health," with their trendy ideas about social medicine.[162]

If nothing else, the failure of the spray-gun war showed the folly of treating malaria as a single disease with a single solution. For when the war fell apart, there were a thousand different reasons why.[163]

In many places today, the spray teams, surveillance workers, and microscopists first put to work on eradicating malaria continue to go through the motions, shadows of the long-dead program. Properly controlling malaria, as opposed to attempting to exterminate it, requires different kinds of workers with different skills, but in many places government leaders feared the political fallout from firing so many eradication workers. So they continue to spray, a little, and collect blood slides, at least a few. But with minimal financing and less oversight, there's no sensible purpose to it.

When chloroquine-resistant *P. falciparum* arrived in Chepo, Panama's Kuna village, the eradication-era workers put in their ghostly appearances: a lone, sporadically paid guy from the country's Vector Control Agency, perhaps, and the occasional sprayman with a canister of insecticide making a few rounds. It's a tepid response that makes little sense given malaria's changing epidemiology or Panama's limited budget. In 1996, Panama collapsed its antimalarial service into the general Health Department, as per WHO's advice. The government has only nineteen cents per capita to spend to tackle malaria. Now when the disease breaks out, they just flex a tired old muscle. The people get bitten, the sprayers head out. The mosquitoes come back. It starts over again.[164]

10. THE SECRET IN THE MOSQUITO

After the ignominious failure of the 1950s-era DDT blitz, malaria disappeared from the headlines. Books on the topic went out of print. Scientists stopped studying the disease; educators stopped teaching it. So completely did malaria vanish from the public mind that many people in the West grew up thinking that there was, literally, no more malaria in this world.

Take Lance Laifer, a hedge-fund manager turned antimalaria organizer. He knew nothing of malaria until he happened to catch a television program on it in 2005. "I didn't know it still existed," he later told *The Wall Street Journal*. "I didn't know it was still killing people. I thought it was eradicated a long time ago. I was just flabbergasted."[1]

In fact, over the course of Laifer's lifetime, the problem of malaria had worsened considerably, especially in Africa.

Although the planet's malarial heartland, sub-Saharan Africa, was excluded from the global malaria-eradication program of the 1950s and 1960s, eradicationists hadn't given up on the continent altogether. In 1960, WHO decided to launch what it called a "pre-eradication" campaign in Africa, a series of demonstration projects and studies that would lead to a viable eradication strategy for the continent. The plan, WHO said, would eradicate malaria from Africa by 1979.[2]

Eradication projects commenced in Liberia, Cameroon, Uganda, and, most extensively, in Nigeria. But it couldn't be done. Even in Nigeria, after six years of near-perfect coverage with proven-effective chemicals, malaria hung on. The evidence appeared irrefutable. Even with the best, most effective tools—and no matter how much money was spent—malaria parasites would always find succor on the African savannah, a stronghold from which to perennially infect and reinfect the rest of the continent.

With eradication off the table, public health experts had little else to recommend to fight *Plasmodium*. The fractious dissolution of the global eradication program had left the antimalaria community underfinanced and roiled with misgivings. The prospects of even reducing malaria seemed challenging, given how compromised both DDT and chloroquine, the two cheapest and most effective antimalaria weapons, had become. Chloroquine-resistant parasites and DDT-resistant mosquitoes, grizzled survivors of the eradication era, lurked across the continent. And DDT in particular had become anathema. The United States banned it in 1972 (by then the chemical industry had moved on to more lucrative and often more toxic insecticides) and geared up to push for an international ban as well. When that happened in 2000, many African governments censored DDT, too, despite a last-minute exclusion for its use against public health threats. Few public health leaders stuck their neck out for the internationally despised chemical.[3]

That's not to say there was nothing that could be done about malaria in Africa. Even without DDT and chloroquine, there were many possible strategies and tools that could tame the scourge, from leveling roads and providing electricity and safe water, to improving housing and installing mosquito-proof window screening, strengthening healthcare systems and heightening public awareness. Mosquito habitats could be minimized, managed, and avoided. Simply paying local clinicians more money, or beefing up a local health clinic, could have helped lighten the malarial burden, as the health policy expert Anne Mills has pointed out.

There wasn't much interest on the part of international anti-malaria financiers in any of this, however. The world's biggest funder of aggressive attacks against disease, the United States, had all but written off the notion of helping African countries with anything. "It is highly unlikely that most African countries will obtain external assistance or investment on anything approaching the scale required for sustained economic development," a CIA report noted in 1965. Nevertheless, as one National Security Council staffer put it at the time, "substantial increases in U.S. foreign assistance expenditures [to Africa] are not envisaged."[4]

African development could have constrained malaria's spread regardless, but over the course of the 1970s and '80s, the forward progress of the continent reversed. The global recession of the 1970s hit newly independent African economies hard, and by the mid-1980s, the World Bank and IMF had taken over the $1.3 trillion in debt they and other developing countries had accrued. The World Bank, which had become the developing world's single largest source of healthcare financing, considered free public health programs a thing of the past, and so required debtor countries to decentralize their health services and encourage privately run clinics and hospitals to sell health care to those consumers willing to pay.[5] In Zaire, more than eighty thousand clinicians and teachers were fired under World Bank and IMF strictures in a single year. In Zambia, within just two years of such programs, infant mortality rose by 25 percent while life expectancy dropped from fifty-four to forty years.[6] In newly crowded African cities, nascent basic services crumbled. Vegetation started to grow in the cracked concrete of urban slums, and residents set out their empty vats and bins to collect rainwater.

Malaria, long known as a disease of the countryside, started to conquer urban slums throughout Africa in the late 1980s.[7] A 1988 flood in the city of Khartoum, for example, set off a malaria outbreak with more than twenty-five thousand cases.[8]

The human immunodeficiency virus exploded in this malarious hotbed, reaching epidemic proportions by the 1980s.[9] AIDS was like a nuclear bomb to malaria's slow poison. For years, infectious disease experts noted the geographic overlap of the two diseases. It wasn't until the late 1990s that scientists started to understand how malaria may have facilitated the spread of HIV, and vice versa.

HIV-positive people are most infective to others when the levels of virus in their bodies are high, which is why HIV treatment that reduces the viral load also slows the virus's spread. With each logarithmic increase in viral load, the probability that an HIV-infected person will transmit the virus during sexual intercourse increases by nearly 250 percent.

Malaria triggers just such spikes. Malaria infection, by inducing HIV to replicate, increases the viral load in HIV-infected people by nearly one log. HIV infection likewise makes its victims more susceptible to malaria. According to mathematical models, since 1980, HIV may have been responsible for 980,000 episodes of malaria, and malaria responsible for more than 8,000 HIV infections, in a single district of Kenya. The global effect of the malevolent partnership between the two pathogens has yet to be mapped.[10]

Overwhelmed with the AIDS crisis, WHO excused ailing African governments from any official responsibility to do something about malaria, calling for the dissolution of government-led antimalaria programs in 1992.[11] An increasingly untameable malaria in Africa became essentially a private affair. Sufferers could seek out a few doses of medication if they wanted to, or find their way to a clinic if they had the cash.

Many years passed before malaria reemerged in the public mind.

Two new antimalarial weapons appeared on the scene. Like DDT and chloroquine, both were easy to use, relatively inexpensive, and relied on the potency of chemicals. One was the ancient Chinese remedy sweet wormwood, which after years of obscurity had been refashioned into cutting-edge artemisinin medications effective against chloroquine-resistant parasites. The other was the humble mosquito bed net, doused with insecticides. In experimental trials in Gambia, researchers found that if meticulously used, insecticide-treated bed nets could slash child mortality from malaria by 20 percent.[12]

Artemisinin drugs and treated nets filled public health officials with hope. "We have enough knowledge, skills and tools," the new head of WHO, Gro Brundtland, proclaimed in 1998, "to launch a new concerted effort" against malaria.[13] The new international antimalaria campaign she sparked, called Roll Back Malaria (RBM), started with the World Bank and various UN agencies and soon grew to include celebrities, corporations, and top NGOs. Hosting conferences, press briefings, concerts, and antimalaria projects, the campaign hoped to inspire a new social movement among donor countries and NGOs to help stanch the bloodshed from malaria.

A new fight against malaria was on. A score of philanthropic, charity, and aid groups formed in RBM's wake, reaching deep into religious, entertainment, sports, and corporate communities. Some of the most powerful people in the world started new NGOs and philanthropic organizations dedicated to saving Africans and others from malaria, from Microsoft founder Bill Gates, former presidents Bill Clinton and George W. Bush, and British prime minister Gordon Brown to U2 front man Bono, celebrated economist Jeffrey Sachs, and News Corporation executive Peter Chernin. Together, they helped increase the annual kitty to fight the disease from a paltry $100 million a year in 1998 to over $1 billion in 2008.[14] Along with the well-regarded Global Fund to Fight Tuberculosis, AIDS and Malaria and the Bill and Melinda Gates Foundation, there was Veto the 'Squito, a youth-led charity, Nothing but Nets, an antimalaria basketball charity, and World Swim Against Malaria.[15] There was the business coalition's Malaria No More, and Laifer's Hedge Funds vs. Malaria, which launched an antimalaria Facebook campaign. Movie stars such as Ashton Kutcher started Twitter campaigns to raise money for malaria.[16] The United States's most popular television program, *American Idol*, devoted a special televised event to raising funds for malaria.[17]

Within ten years of RBM's formation, malaria had gone from being a forgotten, neglected disease to being the "latest in charity gift chic," as *The Guardian* put it in 2007[18]; the "hip way to show you care," *The New York Times* added in 2008[19]; a "cause célèbre," CNN reported.[20]

Many of the charitable individuals, philanthropies, and government agencies who help fund the new antimalaria movement promise far more than they eventually commit. In 2004, for example, though nearly $6 billion had been pledged to Roll Back Malaria, the organization had only $146 million to distribute.[21] In 2009, the Global Fund to Fight AIDS, Tuberculosis and Malaria foresaw a $5 billion shortfall.[22]

What this means is that to maintain its financial base, the new movement must continually court potential donors. This leads to certain difficulties. Malaria is nothing if not a complicated and difficult-to-measure phenomenon. But donors want their charitable dollars to work their magic, not in ten years, not after some other complicated social transformation, but now. If such progress is not demonstrable, attention turns to more pressing matters, and the checks peter out.

For the movement to maintain its momentum, the insoluble problem must become soluble; the complex, simple.

The treated net, they say, is a simple, effective solution that will be readily adopted, so long as sufficient funds are available to buy and distribute them. One of Malaria No More's signature campaigns involves enticing American schoolchildren to raise money to buy treated nets. One child, *The New York Times* reported, built a diorama of an African family in a hut using a pizza box and some Barbie dolls. In the skit she put on, the dolls tucked their little ones into bed with their treated nets. "'She tucks it in, she says, "You're safe now,"'" her mother noted proudly. "'Kids get this in like ninety seconds.'"[23] Which is, of course, the whole point. The disease and its dissolution must become comprehensible to a six-year-old in less than two minutes.

"I have never, ever seen an issue that has greater civic power than malaria," said Malaria No More's John Bridgeland. "The individual literally can step forward, make a contribution, buy a bed net and directly save a life."[24] The treated net is a "simple solution to a devastating problem," CNN noted.[25] By donating ten dollars, you can "buy a mosquito net to save an African child from malaria," *The New York Times* wrote.[26] The Muppets planned a new project to show African children how to use treated nets. "We will save many many lives indeed," an advocate noted. "I just want to stress the extraordinary power of the net," the filmmaker Richard Curtis said at a high-level antimalaria gathering. "For ten dollars people can buy something specific that can save someone's life."[27]

Since ancient times, people have used netting to "catch fish by day," as the fifth-century BC Greek historian Herodotus noted, and "creep under" by night.[28] But twentieth-century malariologists hadn't been particularly enthusiastic about using bed nets to repel malarial mosquitoes. Ronald Ross considered bed nets useful mostly for capturing mosquitoes, not repelling them. He'd make his servants sleep under old nets with a few holes in them in order to collect the engorged *Anopheles* in the net in the morning.[29] Lewis Hackett dismissed bed nets as "a nuisance and an anachronism" that few would use carefully.[30]

After all, many rural peoples in poor countries don't usually sleep on beds per se, opting for mats on the floor. Sleeping within a tent of gauze is hot, too, an added unpleasantness in tropical climes. It's a cultural practice in some countries, but primarily during the months when pest mosquitoes are biting. Elsewhere, families in malarious countries are not unlike my own in India, stringing up the bed nets for the foreigners, but reclining in the open air themselves.

So while it's true that sleeping under a treated net is simple and effective, it is so only in the same way that, say, physicians washing their hands before attending to their patients is simple and effective. A 2009 study in France, for example, revealed that less than half of hospital clinicians followed guidelines on keeping their hands clean, despite the 150-year-old insight that doing so saves the lives of their patients.[31] Just because something is simple doesn't necessarily mean that people will do it.

Reaping the benefits of the treated nets depends on the cooperation of those who use them. And many don't use them as directed, for a variety of complicated social and economic reasons. Anthropologists, reporters, and aid workers have documented how rural Africans handle the nets. They wash them, cleansing them of the insecticides that make them effective. They won't bring their nets to treatment centers for reapplications of insecticide, because they don't want to show their dirty nets outside the home.[32] They refuse to hang the nets over their children. In Zanzibar, as one woman told a local reporter, people felt that the nets "can cause death to children and also cause infertility in women."[33] In Ghana, nets are used to create privacy, so children are considered to not need them. In Gambia, nets are considered expensive items too dear to bestow upon mere children.[34] Elsewhere, the nets are rejected outright. In Ghana, according to malaria expert Philip Adongo, rural people don't like to use insecticides in their homes.[35] In Namibia, people prefer to use the mosquito nets for fishing.[36] And they won't tell distributors about any of these dilemmas. Anthropologists in Ghana, for example, found that "within the local culture it was considered unacceptable to complain about something provided at no charge."[37]

For all these reasons and more, according to a 2003 study, fewer than 17 percent of Africans who received treated nets actually hung them up over their sleeping children.[38]

What's more, the only insecticides deemed safe and effective enough for use on the nets hail from a class of chemicals called pyrethroids. The chemical industry launched pyrethroid insecticides back in the late 1970s,[39] as synthetic versions of the natural insecticide pyrethrum, an extract from the chrysanthemum plant.[40] African farmers along with agriculturalists around the world have been spreading pyrethroid insecticides on their cotton and rice fields for decades.[41] Mosquitoes nurse their young amid the pyrethroid-treated crops, dispatching generations of insects impervious to the toxin. Scientists reported the first *Anopheles gambiae* that had grown insensitive to the killing action of pyrethroids in 1993, five years before Roll Back Malaria launched its pro-net program.[42] Since then, pyrethroid-resistant mosquitoes have turned up across West Africa, and in the Central African Republic, Egypt, Kenya, Mozambique, South Africa, Sudan, and Zimbabwe.[43] The treated nets have thus already started to fail in some places. In a 2005 study in Cameroon, *Anopheles* mosquitoes infected just as many kids using the treated nets as those using untreated ones. The expensive, high-tech insecticide-doused net became just another bit of mesh.

As it turns out, the genes that allow *Anopheles* to circumvent DDT also enable it to rout pyrethroids. For these defiant mosquitoes, no insecticide will suffice, the malariologist Josiane Etang said in 2005. "We need a gun to shoot it so it will die."[44]

While privately, malariologists and others involved in the new antimalaria movement acknowledge the various difficulties associated with treated nets, very little of this gets mentioned publicly in the avalanche of press releases, brochures, and fund-raising appeals their groups issue. Most tend to focus myopically on the number of nets they've distributed—as if this figure were the most compelling one—rather than the number of nets actually used to good effect. "The malaria initiative is a really cool initiative," said former president

George W. Bush at a private 2009 event described by Politico.com. "It's measurable. You get to measure how many nets that have been distributed. It's easy to measure, and it's easy to implement."[45]

Distributing treated nets *is* simple. A single volunteer can distribute hundreds from the back of a motorcycle, and doesn't have to return with more for years. With newer insecticides, the treated nets can repel and poison mosquitoes for up to five years at a time. But equating distribution with use is like counting the bars of soap in the hospital ward rather than the number of clinicians with clean hands. It doesn't tell you much about how many lives are being saved.

It does, however, make the fight against malaria seem straightforward, as do claims that similar battles against malaria have steadily beaten back the disease in the past. Youthful, tall, and photogenic, the Clinton Foundation antimalaria activist Oliver Sabot spends his time jetting around the world advising African ministers of health and global business leaders who want to help the foundation end malaria. At a 2008 meeting, Sabot shows pictures of world maps in which each country that has ever experienced a touch of malaria is darkly shaded, with no regional distinctions. This has the effect of making it seem as if malaria once uniformly blanketed the entire globe save for the poles. All of Australia is shaded, instead of just its northern tip. All of the United States, instead of the bowl-shaped blot malaria actually put upon it. Then, in quick succession, he shows the map as it changes over time, as various countries beat back malaria. In one click, half the shaded countries turn color. Upon the next click, another quarter have turned. In his mythical scenario, all that is needed is one final click.

Cue a new map: Africa, the final frontier. Sabot has pictured lines drawn across the continent, against which malaria can be held at bay, until the line is pushed forward and inward, and all the malaria in the world exists in one tiny puddle.[46] It's a depiction of malaria as a spill that has been methodically wiped away, rather than the tenaciously clinging tick it really is. It gives "an extraordinarily false sense of what we know," remarked the retired WHO scientist Socrates

Litsios, who witnessed the presentation.[47] What it lacks in historical accuracy, though, it makes up for in political appeal.

A similar bias toward political expediency as opposed to accuracy can be seen in the statistically questionable methods that antimalaria organizations often use to track their progress against the disease. Roll Back Malaria, for example, measures the effectiveness of its campaign by comparing the malaria burden in 2000, before their interventions began, to the malaria burden in 2010, ten years later.

Malaria is a naturally fluctuating phenomenon tied to long-term trends in climate, environment, and population movements. Measuring changes in malaria by comparing two distant points, regardless of what happens in between, is less than informative. Your results depend entirely upon what part of the cycle you're on when you start counting. Start from a peak and end on a trough and you can create the illusion of an downward trend. Start from a trough and end on a peak, an upward one. When a delegation of representatives from Myanmar told the RBM organizers that they'd already witnessed precipitous drops in malaria deaths—before the campaign commenced—the organizers' faces dropped, malariologist Andy Spielman says. Malaria had been beaten back, but RBM's ability to claim success had been derailed. "Why couldn't they just wait a year?" Spielman imagined them thinking.[48]

Add to that the fact that during the same period during which RBM said it would slash the malaria death toll in Africa by half,[49] the World Health Organization shifted its statistical estimation techniques, with the result that their estimate of the global malaria burden fell by 50 percent.[50] Half of the world's malaria disappeared, thanks to math. Finally, the most rigorous data is collected only once every five years, and then only during the dry season.[51]

Antimalaria groups tout the flawed numbers, regardless. Not because they have some alternative understanding of their accuracy. Rather, large numbers of distributed nets and rapidly declining malaria numbers create a sense of methodical forward progress, in the five- to ten-year chunks ideal for fund-raising. It fuels the fight against malaria, but it drives the malaria scientists crazy. "It is just not right," fumed Spielman, clenched fists raised. "It is just playing games! . . . This is science, you can't just throw numbers around like that!"[52]

Many of the most persuasive antimalaria leaders today believe that attacking malaria is not just a public health goal—it is a way to attack poverty itself. The leader of this school of thought is Jeffrey Sachs, the economist famous for advising "shock therapy" in the 1980s and author of bestselling books on poverty. Since 2001, when Sachs co-wrote an influential paper detailing the economic burden of malaria, he's argued in prominent magazines and newspapers that antimalaria work should no longer be seen as a public health expense but as an economic investment. Ridding Africa of malaria, Sachs says, will rid Africa of poverty, too.

Malaria is undoubtedly deeply implicated in poverty, as even a seven-year-old child ensconced in a mosquito net can sense, however inchoately. Malaria makes it difficult for farmers to reap their harvests, undermines investment in children, and diverts precious funds toward the purchase of treatments. In Malawi, for example, the average household loses more than three weeks of work to malaria, hemorrhaging over a third of its annual income on the cost of treatment and prevention and in lost workdays.[53] The African continent as a whole loses roughly $12 billion a year due to malaria.[54]

Casting antimalaria work as an investment rather than an expense has undoubtedly broadened the pool of people willing to ante up for treated nets and new drugs. And yet, while Sachs and others have conducted widely cited studies on the correlations between malaria and poverty, none has been able to pinpoint a cause-and-effect relationship. Does malaria cause poverty, as they say, or conversely, is poverty responsible for malaria? If malaria is the trigger, as Sachs maintains, then banishing malaria should be like turning the tap on poverty. But what if it is the other way around, and poverty causes malaria? Then extracting malaria from a community—as in Sri Lanka and Sardinia—could leave scarcity and deprivation essentially intact.

The desire to support economic development is not the only motive driving the new antimalaria movement, though. Many businesses, government leaders, and philanthropists newly drawn to the malaria cause have stumbled upon malaria while pursuing other interests. Corporations want access to Africa's natural resources. Government leaders want to diminish terrorism. Ideologues want to berate political enemies. The list goes on. All have found they can use antimalaria activism to help reach their goals.

Oil companies, for example, have been pursuing new petroleum resources in West Africa—where "oil-filled, undrilled . . . treasures await," as one petro-exploration society put it in 2002—since the late 1980s.[55] Local people in Africa call the companies "oil mosquitoes."[56] The oil hunt inevitably entangled those companies in an expensive fight against African malaria. In 2002, for example, Marathon Oil expanded its natural gas operations into Bioko Island, off the coast of Equatorial Guinea, and was forced to embark on a malaria-control program costing $12 million.[57] ExxonMobil had faced a similar challenge earlier. In the late 1980s and early '90s, while ExxonMobil developed new oil finds in Chad, its mostly foreign workers suffered a malaria rate of 20 percent, which ultimately cost the company around $13.5 million. (Each worker who fell ill with cerebral malaria, for example, had to be evacuated to the Netherlands, at a cost of $100,000 per case.) ExxonMobil eventually launched a $3 million malaria-control program.[58] One in three workers at mining giant Billiton's facility in southern Mozambique suffered malaria, even after the company built a medical clinic, sprayed the construction site, and handed out bed nets. "It was a huge disaster," a spokesperson said. "If we didn't treat malaria we could not operate."[59]

Corporations' efforts aimed at protecting their own workers naturally extend into local communities. Billiton says it could not protect its investment so long as malaria raged in the capital city of

Maputo, just ten miles away, which is why it got involved in a regional antimalaria effort in 2000.[60] While the companies and the local people thus share an interest in taming malaria, there are some notable distinctions. What matters most to the companies, says Spielman, who has consulted for some of them, is not "how many people are dying. What really matters is the entomological inoculation rate," which measures how risky the environment is for outsiders. After five years of spraying insecticides, Marathon's program reduced malaria parasites in local children by just under 50 percent. But sustaining the gain will require a longer financial commitment than the decade or so that oil and gas companies typically invest in new finds.[61] In the case of Bioko Island, international funders have been asked to step into the predicted vacuum.[62]

As a result of these activities, both Marathon and ExxonMobil have become prominent actors in the international antimalaria movement, a role they publicize widely. On World Malaria Day 2007, Marathon took out a large ad in *The New York Times* to describe its efforts to reduce malaria on Bioko Island. The ad pictured the company's director of corporate social responsibility, Adel Chaouch. "It's been a life-changing experience—for me, and especially for the people of Bioko," Chaouch is quoted as saying. "Leading by doing. That's Marathon."[63] In 2008, ExxonMobil sponsored an antimalaria fundraising television program launched by the wildly popular *American Idol*. "Its logo was even branded at the end," noted a commentator for the *Rochester City Newspaper*.[64]

For the United States, support for the oil industry was only one part of the political and economic incentives that propelled the government to devote resources to the antimalaria fight in 2005. President George W. Bush had visited the continent in his first term in office, the first time a U.S. president had done so.[65] In addition, Africa's role in supporting the global terrorist network Al Qaeda—the network maintains a base in Khartoum, Sudan, and its leader, Osama bin Ladin, had called for jihad in Africa—had risen in significance after the terrorist attacks of September 11, 2001.[66]

"There are two reinforcing trends here," one of Bush's aides told *The Washington Post*, describing the motives behind the administration's antimalaria program. "One of them is the upside of foreign policy moralism. Another one is the growing strategic significance of Africa: the conflict with radical Islam, the problem of failed states and terrorism, and the growing importance of Africa on the resource side: oil." Accordingly, the first five African countries targeted by the President's Malaria Initiative included oil-drenched Angola and Equatorial Guinea, along with copper-rich Zambia.[67]

For free-market conservatives, supporting the antimalaria movement helped score points in ideological wars. They'd long battled the environmental lobby's push for more stringent environmental regulations. Under the theory that the enemy's enemy is a friend, free marketeers have rushed to defend environmentalists' totemic antihero DDT. The free-market economist Roger Bate of the conservative American Enterprise Institute, for example, is one of the most vocal defenders of DDT, which he lauds as "the single most valuable chemical ever synthesized to prevent disease."[68] Africa Fighting Malaria, the group Bate founded, is dedicated to promoting the use of DDT against malaria.

Their message—that by maligning DDT, environmentalists have the blood of malaria victims on their hands—has spread widely. "Banning DDT killed more people than Hitler," the novelist Michael Crichton wrote in 2004. "And the environmental movement pushed hard for it."[69] Crichton aired his views when asked to testify at a congressional hearing on malaria.[70] DDT is "the best thing in our arsenal," the malaria activist Lance Laifer told *The Wall Street Journal* in 2006. "We need to have people walking around with signs that say, 'DDT saves lives, environmentalists take lives.'"[71] In 2007, Republican senator Tom Coburn blocked bills to honor the anti-DDT crusader Rachel Carson. He explained by linking his website to one called Rachel Was Wrong, which states that millions of people suffer malaria "because one person sounded a false alarm . . . Rachel Carson."[72]

President George W. Bush, often accused by the environmental lobby of gutting environmental regulations, pointedly rehabilitated DDT in his antimalaria program. He publicly announced the pro-DDT program at the Hudson Institute, a neoconservative, antienvironmentalist think tank.[73]

The World Health Organization supported, managed, and oversaw the previous global movement against malaria. Critics may argue over how expertly it performed the task, but however one judges its technical leadership, its authority was both sanctioned by and accountable to the international community, that is, the 193-member nations of the UN.

The new global movement against malaria, in contrast, is led by private interests. Malariologists joke that the industrialized country alliance G8 should really be called the G9, to include the Gates Foundation.[74] Disbursing more than $9 billion for global health research between 1998 and 2008,[75] the foundation—not public health authorities—sets the agenda in antimalaria research. "It's true," remarked the malariologist Brian Greenwood, "we are all doing what Gates says."[76]

As a private entity, the Gates Foundation is not beholden to governments or international agencies. When push comes to shove, the foundation can even eclipse the public health authority of the World Health Organization.

Take, for example, an antimalarial strategy called intermittent preventive therapy for infants (IPTI), which calls for sporadic doses of preventive drugs for malaria. The World Health Organization routinely reviews scientific evidence on new methods to provide well-regarded guidance to public health authorities around the world. Public health agencies are not required to abide by WHO's recommendations, but most do: it's considered the standard of care. WHO reviewed the research on IPTI in 2007. Its scientific committee decided not to recommend the therapy's use in antimalaria programs, because it didn't alter mortality and it risked some serious side effects.[77]

The Gates Foundation had funded a host of research on IPTI, however, and felt differently—and so took the unusual step of asking the National Academy of Sciences' Institute of Medicine to draw up another review to compete with and possibly undermine WHO's.[78] Indeed, the Institute of Medicine's review, while agreeing with all the objections WHO outlined, concluded that IPTI was worthwhile, nevertheless.[79]

The World Health Organization's Arata Kochi, a fierce freethinker who took the helm of WHO's malaria program in 2006, was not pleased.[80] In a leaked memo he complained that the Gates researchers were becoming a "cartel" of groupthink. "Each has a vested interest to safeguard the work of others," he wrote. "The result is that obtaining an independent review of scientific evidence . . . is becoming increasingly difficult . . . and could have implicitly dangerous consequences."[81]

The public conflict with the Gates Foundation proved the last public foray for Kochi. According to insiders, he was put on "gardening leave." Although his name still appeared on WHO's website as director of the malaria program, the outspoken director has been conspicuously silent and absent from malaria meetings since 2008.[82]

Expert opinion has likewise been arrayed against the Gates Foundation's stance that the antimalaria movement should make complete eradication of malaria its goal, rather than just attempting to hold the scourge in check. Most malaria experts agree that with more resources, malaria could be eliminated from marginal areas, but that elsewhere, nothing has really changed from the 1950s. All the problems that stymied such ambitions in the past—*Anopheles gambiae*'s tenacity, population movements, resistance to insecticides and drugs, lack of community participation, poor statistics and worse surveillance, and persistent poverty—remain.

In the hallways of malaria meetings and in private conversations, the grumbling has been audible. "I'm appalled . . . They are making all the same mistakes again," one malariologist said. "It's amazing how we don't learn about our own history, isn't it?" another remarked.[83] "The barbarians have taken over," explained another. "The people who don't really know what they are talking about."[84] "Go along with it if you want to get funded," a malariologist said to *The New York Times*. But don't sign on to anything unless eradication is tied to a date like 2050, he said, "or far enough in the future so that none of us can be held accountable."[85]

Few have been brave enough to air their misgivings publicly.

Bill and Melinda Gates announced eradication as the new goal for antimalaria work in late 2007, at a private gathering. Kochi would almost certainly have scoffed at the notion, but the politically savvy WHO director-general Margaret Chan smoothly agreed. "I dare you to come along with us," she told the crowd of skeptical malariologists.[86] Roll Back Malaria, the United Nations, and others quickly signed on, issuing reports and holding press conferences on the new goal.[87]

As more money poured in, fund-raisers and donors started to act as if the hard part of the job had already been done. Said the rock star Bono, a prominent supporter and fund-raiser, at a gala antimalaria event in 2008:

> I'd like to say that I'm not here as a rock star. Really I'm here as a fan. And I'm a fan of Malaria No More, what you two gentlemen have done is extraordinary. I'm a great fan of Africa, in particular. These leaders, incredible. I'm in their fan club. I'm a great fan of the physicians and the scientists who gathered on this problem. Bill Gates. He's a rock star. Jeffrey Sachs, all the people who have ganged up on the problem. People in Red who have campaigned for Global Fund money, it started with AIDS but now it's malaria. It just shows the momentum. It just shows what's possible when you match leadership with funding, a strategic plan. So I'm just going to shut up with that. And just say, what's the next disease? Pneumococcal? Rotavirus? Because, uh, you know this malaria thing is extraordinary and it just shows what else we can do.[88]

• • •

The shameful resurgence of malaria in the 1980s and '90s has, for now, been reversed. After a decade of effort, by 2008, sixty-seven countries suffering endemic falciparum malaria had formally adopted artemisinin combination medications as their first-line remedy for malaria, including forty-one of Africa's fifty-four countries.[89] Twenty million of Africa's one hundred and ten million children under the age of five sleep under treated nets.[90] Grants for scientific research on malaria have created a new generation of high-tech malariologists, who've brought experimental malaria vaccines into late-phase clinical trials.

The antimalaria movement may use hype, suffer conflicts of interest, and have a lack of accountability, but despite this it deserves credit. And yet, the uncomfortable truth is that ending malaria over the long term will require much more difficult social and economic adjustments in African communities, just as it has elsewhere. Infrastructure will have to improve. Settlement patterns and housing styles will have to change. Education and healthcare systems will have to be built.

Antimalaria activists know this. But it is not possible for them alone to transform African economies and cultures. The best they can do is offer partial, short-term solutions. That is, in the meantime, they can blanket the continent with treated nets and better drugs. So long as the charitable dollars keep flowing, lives will be saved—at least for now.

After all, the perfect need not be the enemy of the good. The question is how the short-term solutions impact the prospects for the long-term ones. Usually, something good today doesn't reduce the probability of something better tomorrow. But in malaria, it can. When DDT was touted as a quick win, for example, and when donors promised the imminent arrival of a malaria vaccine, political will and financing for malaria research and other forms of malaria control fell by the wayside. Promised easy victories, political leaders lose the will to fight the long-term battle. And if the short-term solutions prove successful but are not maintained, malaria could resurge, just as it did in Sri Lanka in the wake of last century's failed global eradication blitz.

This conflict over short-term solutions and long-term sustainability has yet to be adequately resolved. The U.S. antimalaria program, government malariologist Thomas Ritchie says, "is pouring obscene amounts of money" into quick fixes against African malaria, but it is spending little on supporting local antimalaria leadership or building antimalaria infrastructure. Plenty of African clinicians, scientists, and community leaders are dedicated to taming malaria, Ritchie says, but when the world's richest country decides to help, "they give these people nothing, not a cent!"[91]

Donors such as the Global Fund offer two-year grants for countries to stock expensive new antimalarial drugs, but leave them with few options when the grants run out. "You cannot make public health policy based on two-year grants, however much money you are being given," one critic complained to *The East African*. "What will happen when the same donors accuse us of corruption and withdraw funding? . . . We are essentially making a donor-supported treatment that we cannot afford into the cornerstone of our malaria treatment."[92]

The conflict plays out in heated debates at international malaria meetings. At one, an official from the Nigerian Ministry of Health became engrossed in a long argument with a representative from the drug giant Sanofi-Aventis, which was at the time the sole provider of WHO-recommended ACT drugs. Finally she turned to me. "Write it in your paper," she commanded. "We need to build African capacity to make treated nets and ACTs. That is the only way we can solve malaria. They don't want to do technology transfer," she said, motioning to the drug company rep. "They just want us to buy, buy, buy!"[93]

Although it publicly recognizes the need to build infrastructure in endemic countries, Roll Back Malaria has also stated that tackling the disease cannot be the responsibility of local governments. "If malaria control is left to governments to plan and execute," RBM wrote, "malaria will not be controlled."[94] Which is, of course, exactly backward. It is the *only* way malaria will be controlled. And malaria-endemic societies have proven this over and over again, from when the Italians distributed quinine to their populace—and built the schools and clinics and roads they needed in order to do it—to when Malawi banned the sale of chloroquine and rid the country of chloroquine-resistant parasites.[95]

Somehow antimalaria work must unleash the technology, political will, and infrastructure in malaria-plagued countries to hold the line and sustain hard-won gains. One way or another, the schools, roads, clinics, secure housing, and good governance that enable regular prevention and prompt treatment must be built. Otherwise, the cycle of depression and resurgence will begin anew; malaria will win, as it always has. "You can do a lot of good with bed nets, with spraying," says malariologist Tom McCutchan, "but in the end, you have got to give power to the people who are at risk."[96]

While we debate, and argue, and haphazardly collect our strength to fight malaria, the parasite refines its plague upon us. Unlike us, *Plasmodium* does not reenact failed strategies and weak defenses, its historical memory shot. The evolution of its predation is progressive, methodical, probing.

Plasmodium may have evolved a fifth species to prey upon humankind. In 2008, researchers found that more than a quarter of a sample of one thousand malaria patients in Malaysia harbored something altogether unexpected: *Plasmodium knowlesi*, a parasite previously believed to be confined to monkeys. As booming, tree-felling human populations increasingly intrude into monkey habitat, experts suspect, they've offered themselves as a new blood source for *P. knowlesi* to exploit. The parasite has already been found in humans in Thailand and China. Whether it will make its rounds into the rest of the malarious world remains unknown. For now, its victims must hope for quick diagnosis and prompt treatment. With the shortest life cycle of any malaria species, *P. knowlesi* can unleash tremendous masses of parasites rapidly.[97]

Chloroquine-resistant falciparum parasites have arrived on the doorstep of the global economy. As drug traffickers and others ply the Caribbean's cerulean waters on their way to the Panama Canal, they stop at the remote beaches that fringe the Panamanian coast, where Kuna communities live much as they have for centuries, untouched by road or rail. When an especially virulent strain of malaria broke out there in 2003, the Panamanian authorities couldn't do much about it. The dissolution of the global eradication campaign and the political neglect of malaria that followed had seen spending on mosquito control in Panama drop from $1.20 per capita per year to just 19 cents.[98]

In 2005, the coastal Kuna paddled their dugout canoes through miles of jungle until they arrived at Chepo, just outside Panama City, for a meeting of Kuna leaders. They strung their hammocks alongside their Kuna brethren who lived there. Every night, Chepo's plentiful mosquitoes feasted on their blood, and then flitted over to the next hammock and bit the locals, too. In the morning, the Chepo Kuna, especially the young ones, who are eager to abandon the old ways, went off to their jobs in the pizza parlors of Panama City, where their warm bodies jostled with those of the tourists, from Michigan and New York and Essex and Rome, discharged from the cruise ships anchored in the canal.

The entire economy, it is said, would have to break down in order for malaria to resettle in developed nations such as the United States. And yet mosquito-borne West Nile virus and Japanese encephalitis have spread unchecked. In 2002, California had a single case of West Nile virus; in 2003, there were three, according to the Centers for Disease Control. By 2004, there were 779 cases nationwide; in 2005, 873. In 2008, there were more than 1,300.[99] The economy survives, despite it.

The U.S. economy tolerates, too, those pockets of humid, neglected anarchy where *Plasmodium* builds its strongholds, such as the drowned cities of the South, deprived of electricity and order in the wake of the 2005 hurricanes. Malaria parasites continually shower upon the nation. Between 2005 and 2006 more than three thousand people in the United States fell ill with malaria picked up from West Africa, Asia, and elsewhere.[100] Every now and again, the local mosquitoes start to transmit the parasites to people who've never broached a U.S. border. Between 1957 and 1994, American mosquitoes infected seventy-four people in the United States with malaria.[101]

It wouldn't take much for the malarial mosquitoes of Europe to start transmitting the parasite once again. Perhaps if Russia cut off Europe's natural gas supply, as it did in the winter of 2006, rendering a blackout that might stall the water pumps. Or if a distracting health emergency occurred, like the 2003 heat wave that killed more than thirty thousand Europeans over the course of a single season.

The local *Anopheles* vectors, in Europe as in North America, are as abundant as they ever were. And every couple of years, the mosquitoes pick up some parasites and start to bite the locals.[102]

Their warm blood beckons.

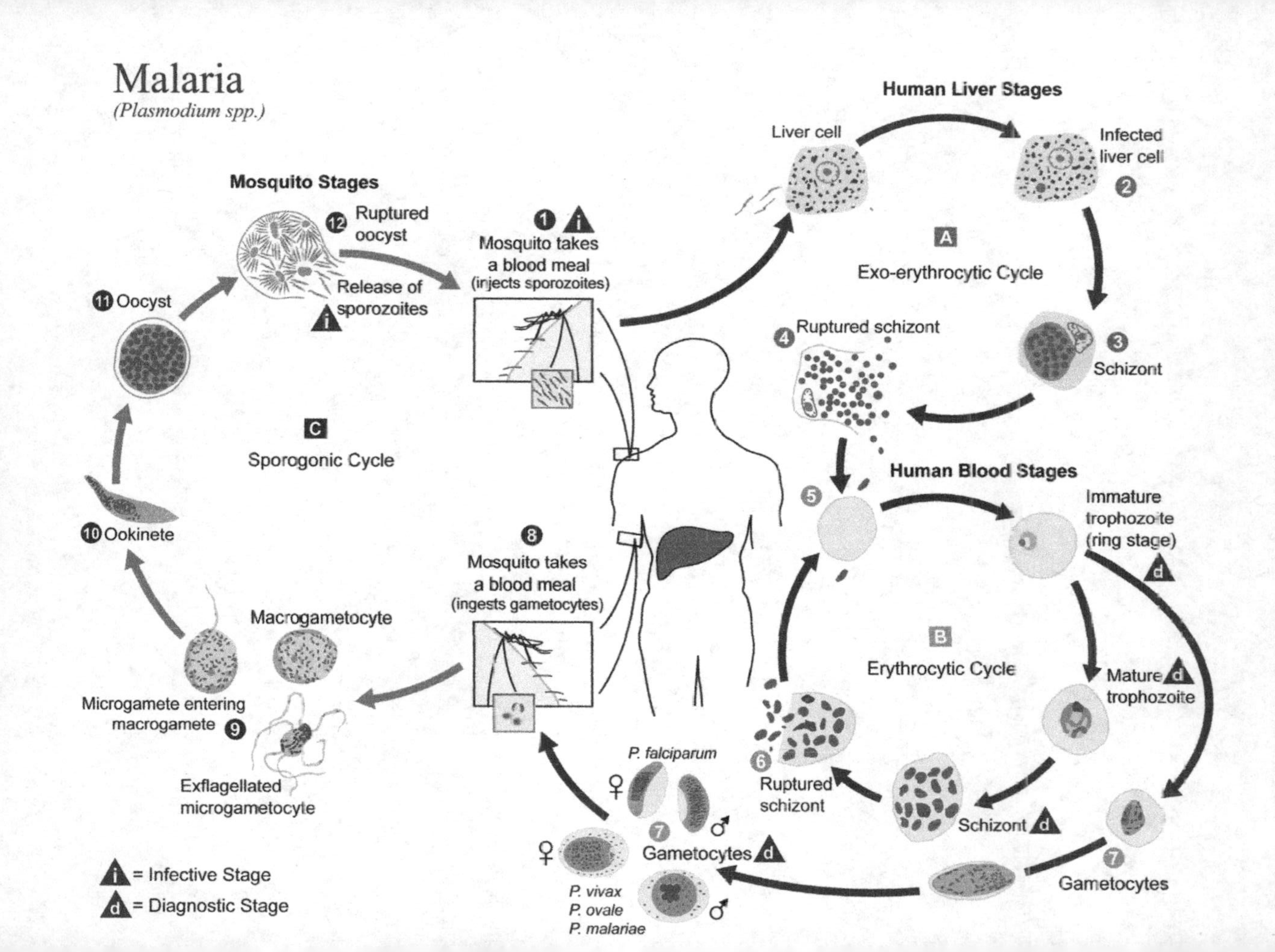

Malaria
(Plasmodium spp.)
Mosquito Stages
12 Ruptured oocyst
Release of sporozoites
11 Oocyst
C
Sporogonic Cycle
10 Ookinete
Macrogametocyte
Microgamete entering macrogamete 9
Exflagellated microgametocyte
1 Mosquito takes a blood meal (injects sporozoites)
8 Mosquito takes a blood meal (ingests gametocytes)
Human Liver Stages
Liver cell
Infected liver cell 2
A
Exo-erythrocytic Cycle
3 Schizont
4 Ruptured schizont
Human Blood Stages
5
Immature trophozoite (ring stage)
B
Erythrocytic Cycle
Mature trophozoite
Schizont
6 Ruptured schizont
7 Gametocytes
P. falciparum
7 Gametocytes
P. vivax
P. ovale
P. malariae
i = Infective Stage
d = Diagnostic Stage

NOTES

1. MALARIA AT OUR DOORSTEP

1. www.stratfor.com/global_market_brief_panama_canal_expansion.
2. Pan American Health Organization, "Malaria in Panama, 1998–2004: Time Series Epidemiological Data from 1998 to 2004."
3. Médecins Sans Frontières, "ACT NOW to Get Malaria Treatment That Works to Africa," April 2003.
4. G. Sabatinelli et al., "Malaria in the WHO European Region," *Euro Surveillance* 6, no. 4 (April 2001): 61–65.
5. World Health Organization, *World Malaria Report 2005*, available at www.rbm.who.int/wmr2005/html/exsummary_en.htm.
6. Sabatinelli et al., "Malaria in the WHO European Region."

2. BIRTH OF A KILLER

1. Interview with Themba Mzilahowa, medical entomologist, Blantyre, Malawi, February 20, 2007.
2. Nicholas A. V. Beare et al., "Malarial Retinopathy: A Newly Established Diagnostic Sign in Severe Malaria," *American Journal of Tropical Medicine and Hygiene* 75, no. 5 (2006): 790–97.
3. Correspondence with Terrie Taylor, March 4, 2007.
4. Interview with Terrie Taylor, February 19–21, 2007.
5. Roy Porter, *The Greatest Benefit to Mankind* (New York: W. W. Norton, 1997), 25.

6. Andrew Spielman and Michael D'Antonio, *Mosquito: A Natural History of Our Most Persistent and Deadly Foe* (New York: Hyperion, 2001), 44–45.
7. Richard Carter and Kamini Mendis, "Evolutionary and Historical Aspects of the Burden of Malaria," *Clinical Microbiology Reviews* 15, no. 4 (October 2002): 579.
8. Carl Zimmer, *Parasite Rex: Inside the Bizarre World of Nature's Most Dangerous Creatures* (New York: Touchstone, 2000), 17.
9. Ibid., 17–18.
10. David J. Marcogliese and Judith Price, "The Paradox of Parasites," *Global Biodiversity* 3 (1997): 7–15.
11. "Herbicide Hope for Malaria," BBC News, January 31, 2003.
12. Graeme O'Neill, "Pathways to Destruction," *The Bulletin*, February 12, 2003.
13. Carter and Mendis, "Evolutionary and Historical Aspects of the Burden of Malaria," 564–94.
14. Lewis W. Hackett, *Malaria in Europe: An Ecological Study* (London: Oxford University Press, 1937), 201.
15. Yuemei Dong et al., "*Anopheles gambiae* Immune Responses to Human and Rodent *Plasmodium* Parasite Species," *PLoS Pathogens* 2, no. 6 (June 2006): e52.
16. R. E. Sinden et al., "Mosquito-Malaria Interactions: A Reappraisal of the Concepts of Susceptibility and Refractoriness," *Insect Biochemistry and Molecular Biology* 34 (2004): 625–29.
17. David A. Warrell and Herbert M. Gilles, eds. *Essential Malariology*, 4th ed. (London: Hodder Arnold, 2002), 59.
18. Angelika Sturm et al., "Manipulation of Host Hepatocytes by the Malaria Parasite for Delivery into Liver Sinusoids," *Science* 313 (2006): 1287–90.
19. Dominic P. Kwiatkowski, "How Malaria Has Affected the Human Genome and What Human Genetics Can Teach Us About Malaria," *American Journal of Human Genetics* 77 (2005): 171–90.
20. Spielman and D'Antonio, *Mosquito*, 15.
21. Ibid., 15–16.
22. Warrell and Gilles, eds. *Essential Malariology*, 12–13.
23. Robert Sallares, *Malaria and Rome: A History of Malaria in Ancient Italy* (Oxford: Oxford University Press, 2002), 16.
24. Robert A. Anderson et al., "The Effect of *Plasmodium yoelii nigeriensis* Infection on the Feeding Persistence of *Anopheles stephensi* Liston Throughout the Sporogonic Cycle," *Proceedings: Biological Sciences* 266 (September 7, 1999): 1729–33.
25. Jacob C. Koella et al., "The Malaria Parasite, *Plasmodium falciparum*, Increases the Frequency of Multiple Feeding of Its Mosquito Vector, *Anopheles gambiae*," *Proceedings of the Royal Society of London* 265 (1998): 763–68.

26. Anthony James, "Blocking Malaria Parasite Invasion of Mosquito Salivary Glands," *Journal of Experimental Biology* 206 (2003): 3817–21.
27. Heather Ferguson and Andrew F. Read, "Why Is the Effect of Malaria Parasites on Mosquito Survival Still Unresolved?" *Trends in Parasitology* 18, no. 6 (June 2002): 256–61.
28. Renaud Lacroix et al., "Malaria Infection Increases Attractiveness of Humans to Mosquitoes," *PLoS Biology* 3, no. 9 (September 2005): e298.
29. Kevin Graham, "Rare Gene Pairing Lethal to Boy," *St. Petersburg Times*, August 23, 2006; D. J. Weatherall and J. B. Clegg, "Inherited Haemoglobin Disorders: An Increasing Global Health Problem," *Bulletin of the World Health Organization* 79, no. 8 (2001): 704.
30. J. D. Smyth, *Introduction to Animal Parasitology*, 3rd ed. (Cambridge, UK: Cambridge University Press, 1994), 126–35.
31. Sallares, *Malaria and Rome*, 12.
32. Ibid.
33. Warrell and Gilles, eds., *Essential Malariology*, 26–27.
34. Ibid., 24–25.
35. Carter and Mendis, "Evolutionary and Historical Aspects of the Burden of Malaria," 564–94.
36. Ibid.
37. Sallares, *Malaria and Rome*, 151, citing Garnham 1966.
38. Warrell and Gilles, eds., *Essential Malariology*, 24–25.
39. Carter and Mendis, "Evolutionary and Historical Aspects of the Burden of Malaria," 564–94.
40. Ibid.
41. Sir Malcolm Watson, *African Highway: The Battle for Health in Central Africa* (London: John Murray Publishers, 1953), 232.
42. Nina L. Etkin, "The Co-evolution of People, Plants, and Parasites: Biological and Cultural Adaptations to Malaria," *Proceedings of the Nutrition Society* 62 (2003): 311–17; James L. A. Webb, "Malaria and the Peopling of Early Tropical Africa," *Journal of World History* 16, no. 3 (2005): 269–91.
43. Francisco J. Ayala and Mario Coluzzi, "Chromosome Speciation: Humans, *Drosophila*, and Mosquitoes," *PNAS* 102, suppl. 1 (May 3, 2005), 6535–42.
44. Mario Coluzzi, "The Clay Feet of the Malaria Giant and Its African Roots: Hypotheses and Inferences About Origin, Spread, and Control of *Plasmodium falciparum*," *Parassitologia* 41 (1999): 277–83.
45. Ayala and Coluzzi, "Chromosome Speciation," 6535–42.
46. Sallares, *Malaria and Rome*, 25.
47. Institute of Medicine, *Saving Lives, Buying Time: Economics of Malaria Drugs in an Age of Resistance* (Washington, D.C.: National Academies Press, 2004), 144.

48. Kwiatkowski, "How Malaria Has Affected the Human Genome and What Human Genetics Can Teach Us About Malaria," 171–90.
49. Carter and Mendis, "Evolutionary and Historical Aspects of the Burden of Malaria," 564–94.
50. Kwiatkowski, "How Malaria Has Affected the Human Genome," 171–90.
51. Weatherall and Clegg, "Inherited Haemoglobin Disorders," 704–12.
52. Siske S. Struik and Eleanor M. Riley, "Does Malaria Suffer from Lack of Memory?" *Immunological Reviews* 201 (2004): 268–90.
53. Hackett, *Malaria in Europe*, 172.
54. Interview with Dr. Simon Glover, QEH, Blantyre, Malawi, February 21, 2007; Beare et al., "Malarial Retinopathy," 790–97.
55. Kwiatkowski, "How Malaria Has Affected the Human Genome," 171–90.
56. Ibid.
57. C. Dobano et al., "Expression of Merozoite Surface Protein Markers by *Plasmodium falciparum*–infected Erythrocytes in Peripheral Blood and Tissues of Children with Fatal Malaria," *Infection and Immunity* 75, no. 2 (February 2007): 643–52.
58. Warrell and Gilles, eds., *Essential Malariology*, 206.
59. Estimate is of one million deaths from malaria every year, 700,000 to 900,000 of which occur among African children under the age of five.

3. SWEPT IN MALARIA'S CURRENT

1. "The Kingdom of Thirst," *New York Times*, March 27, 1884.
2. Address by Peter Asoka, Fourth MIM Pan-African Malaria Conference, November 2005, Yaoundé, Cameroon.
3. James L. A. Webb, "Malaria and the Peopling of Early Tropical Africa," *Journal of World History* 16, no. 3 (2005): 269–91.
4. R. L. Miller et al., "Diagnosis of *Plasmodium falciparum* Infections in Mummies Using the Rapid Manual *Para*Sight-F Test," *Transactions of the Royal Society of Tropical Medicine and Hygiene* 88 (1994): 31–32.
5. Leonard Jan Bruce-Chwatt and Julian de Zulueta, *The Rise and Fall of Malaria in Europe: A Historico-epidemiological Study* (Oxford: Oxford University Press/Regional Office for Europe of the World Health Organization, 1980), 17.
6. Robert S. Desowitz, *The Malaria Capers: More Tales of Parasites and People, Research and Reality* (New York: W. W. Norton, 1991), 150.
7. Richard Carter and Kamini Mendis, "Evolutionary and Historical Aspects of the Burden of Malaria," *Clinical Microbiology Reviews* 15, no. 4 (October

2002): 564–94, quoting from H. E. Sigerist, *A History of Medicine, Volume 1: Primitive and Archaic Medicine* (New York: Oxford University Press, 1951).

8. Herbert S. Klein, *The Atlantic Slave Trade* (Cambridge, UK: Cambridge University Press, 1999), 68.
9. Philip D. Curtin, *Disease and Empire: The Health of European Troops in the Conquest of Africa* (Cambridge, UK: Cambridge University Press, 1998), 21.
10. Letter from David Livingstone to Dr. James Ormiston McWilliam, November 28, 1860, available at www.livingstoneonline.ucl.ac.uk, published in *Transactions of the Epidemiological Society of London*, 1860.
11. Klein, *The Atlantic Slave Trade*, 59.
12. Philip D. Curtin, "Epidemiology and the Slave Trade," *Political Science Quarterly* 83, no. 2 (June 1968): 190–216; Curtin, *Disease and Empire*, 1.
13. Curtin, *Disease and Empire*, 3.
14. Ann Vileisis, *Discovering the Unknown Landscape: A History of America's Wetlands* (Washington, D.C.: Island Press), 4.
15. Ibid., 16.
16. W. V. King and G. H. Bradley, "Distribution of the Nearctic Species of Anopheles" and "Bionomics and Ecology of Nearctic Anopheles," in Forest Ray Moulton, ed., *A Symposium on Human Malaria with Special Reference to North America and the Caribbean Region* (Washington, D.C.: American Association for the Advancement of Science, 1941), 71–87.
17. Jon Kukla, "Kentish Agues and American Distempers: The Transmission of Malaria from England to Virginia in the Seventeenth Century," *Southern Studies* 25, no. 2 (Summer 1986): 135–47.
18. Quoted in Kukla, "Kentish Agues and American Distempers," 135–47.
19. See Thomas J. Wertenbaker, *Virginia Under the Stuarts, 1607–1988* (Princeton, N.J.: Princeton University Press, 1914), 11; Darrett B. Rutman and Anita H. Rutman, "Of Agues and Fevers: Malaria in the Early Chesapeake," *The William and Mary Quarterly* 33, no. 1 (January 1976): 31–60; Alan Taylor, *American Colonies: The Settling of North America* (New York: Penguin, 2001), 130–31. Margaret Humphreys agrees with Wyndham Blanton and Carville Earle, who argue that the deadly fevers that afflicted the Jamestown colonists were probably typhoid, not malaria, because the colonists were not nonimmune to *vivax*, and there was no source of *falciparum*—Margaret Humphreys, *Malaria: Poverty, Race, and Public Health in the United States* (Baltimore: Johns Hopkins University Press, 2001), 24.
20. Quoted in Rutman and Rutman, "Of Agues and Fevers," 31–60.
21. Taylor, *American Colonies*, 145.
22. Ibid., 147.

23. Kukla, "Kentish Agues and American Distempers," 135–47.
24. Mary J. Dobson, "Mortality Gradients and Disease Exchanges: Comparisons from Old England and Colonial America," *Social History of Medicine* 2, no. 3 (December 1989): 259–97.
25. Oliver Wendell Holmes, *Boylston Prize Dissertations for the Years 1836 and 1837* (Boston: Charles C. Little and James Brown, 1838), 11–12.
26. Dobson, "Mortality Gradients and Disease Exchanges," 259–97.
27. Klein, *The Atlantic Slave Trade*, 21, 27, 28, and Curtin, "Epidemiology and the Slave Trade," 190–216.
28. Klein, *The Atlantic Slave Trade*, 2.
29. Curtin, "Epidemiology and the Slave Trade," 190–216.
30. Klein, *The Atlantic Slave Trade*, 10.
31. Ibid., 72, 120–26.
32. Ibid., 91.
33. Ibid., 77, 122, 125, 152.
34. Alexander Falconbridge, *An Account of the Slave Trade on the Coast of Africa* (London: J. Phillips, 1788), 11, 51.
35. Klein, *The Atlantic Slave Trade*, 81, 125.
36. St. Julien Ravenel Childs, *Malaria and Colonization in the Carolina Low Country 1526–1696* (Baltimore: Johns Hopkins University Press, 1940), 28.
37. Quoted in Karen Ordahl Kupperman, "Fear of Hot Climates in the Anglo-American Colonial Experience," *William and Mary Quarterly* 41, no. 2 (April 1984): 213–40.
38. James Stevens Simmons, *Malaria in Panama* (Baltimore: Johns Hopkins University Press, 1939), 6.
39. Ibid., 4.
40. John Prebble, *The Darién Disaster: A Scots Colony in the New World, 1698–1700* (New York: Holt, Rinehart and Winston, 1968), 76.
41. Simmons, *Malaria in Panama*, 6.
42. Ibid., 6–7.
43. Ibid., 4.
44. Ignacio J. Gallup-Diaz, *The Door of the Seas and Key to the Universe* (New York: Columbia University Press, 2000).
45. Taylor, *American Colonies*, 217.
46. Ravenel Childs, *Malaria and Colonization*, 245.
47. Ibid., 231.
48. Ibid., 221.
49. Peter H. Wood, *Black Majority: Negroes in Colonial South Carolina from 1670 Through the Stono Rebellion* (New York: Alfred A. Knopf, 1975), 67.
50. Ibid.

51. Fiammetta Rocco, *The Miraculous Fever-Tree: Malaria and the Quest for a Cure That Changed the World* (New York: HarperCollins, 2003), 171–74.
52. H. Roy Merrens and George D. Terry, "Dying in Paradise: Malaria, Mortality, and the Perceptual Environment in Colonial South Carolina," *Journal of Southern History* 50, no. 4 (November 1984): 542.
53. Jill Dubisch, "Low Country Fevers: Cultural Adaptations to Malaria in Antebellum South Carolina," *Social Science and Medicine* 21, no. 6 (1985): 641–49.
54. Todd L. Savitt, "Slave Health," in Todd L. Savitt and James Harvey Young, eds., *Disease and Distinctiveness in the American South* (Knoxville: University of Tennessee Press, 1988), 124–25.
55. Quoted in Wood, *Black Majority*, 83.
56. Quoted in Dubisch, "Low Country Fevers," 641–49.
57. See "Prevalence of the Sickle Cell Trait in Adults of Charlestown County, S.C.: An Epidemiological Study," *Archives of Environmental Health* 17 (1968): 891–98, quoted in Wood, *Black Majority*, 89.
58. Curtin, "Epidemiology and the Slave Trade," 190–216.
59. Taylor, *American Colonies*, 231.
60. Prebble, *The Darién Disaster*, 17–18.
61. Ibid., 12.
62. Ibid., 63.
63. Wafer had also helpfully aquired antimalarial cinchona bark in northern Chile in the 1680s. James L. A. Webb, *Humanity's Burden: A Global History of Malaria* (New York: Cambridge University Press, 2009), 96. Also Prebble, *The Darién Disaster*, 68–69.
64. See www.bbc.co.uk/weather/features/understanding/scotland_01.shtml.
65. Prebble, *The Darién Disaster*, 42.
66. Ibid., 91.
67. Quoted in Dennis R. Hidalgo, "To Get Rich for Our Homeland: The Company of Scotland and the Colonization of the Darién," *Colonial Latin American Historical Review* 10, no. 3 (Summer/Verano 2001): 311–50.
68. Prebble, *The Darién Disaster*, 61, 65, 80, 97–100.
69. Ibid., 120–28.
70. Klein, *The Atlantic Slave Trade*, 134.
71. They sent just one hundred soldiers, out of four hundred who left Scotland. Prebble, *The Darién Disaster*, 128–44, 151–72, 189.
72. Gallup-Diaz, *The Door of the Seas and Key to the Universe*.
73. Hidalgo, "To Get Rich for Our Homeland."
74. From National Archives of Scotland, GD406/1/4372.
75. Prebble, *The Darién Disaster*, 176, 182, 198.

76. Letter from George Douglas, April 10, 1699. From National Archives of Scotland, GD446/39/16.
77. From National Archives of Scotland, GD406/1/4372.
78. Prebble, *The Darién Disaster*, 204.
79. Ibid., 200–46.
80. Letter from Alexander Shields, February 2, 1700. From Registrar General for Scotland, OPR453/9, p. 139, cited in National Archives of Scotland, "The Darién Adventure," text to an exhibition, 1998–1999.
81. Prebble, *The Darién Disaster*, 247, 255.
82. Christopher Storrs, "Disaster at Darién (1698–1700)? The Persistence of Spanish Imperial Power on the Eve of the Demise of the Spanish Habsburgs," *European History Quarterly* 29, no. 1 (1999): 5–37.
83. Prebble, *The Darién Disaster*, 269–307.
84. Mike Ibeji, *The Darién Venture*, BBCi History, January 5, 2001, available at www.bbc.co.uk/history/state/nations/scotland_darien_01.htm.
85. Prebble, *The Darién Disaster*, 311.
86. Andrew Spielman lecture, "Malaria and Human Affairs," course, Harvard University, March 2, 2006.
87. James O. Breeden, "Disease as a Factor in Southern Distinctiveness," in Savitt and Young, eds., *Disease and Distinctiveness in the American South*, 3.
88. Taylor, *American Colonies*, 154.
89. Humphreys, *Malaria: Poverty, Race, and Public Health in the United States*, 25.
90. Taylor, *American Colonies*, 154.
91. Samuel A. Cartwright, "Report on the Diseases and Physical Peculiarities of the Negro Race," *New Orleans Medical and Surgical Journal* (May 1851): 694.
92. Quoted in J.D.B. De Bow, *The Industrial Resources, Etc., of the Southern and Western States: Embracing a View of Their Commerce, Agriculture, Manufactures, Internal Improvements; Slave and Free Labor, Slavery Institutions, Products, etc., of the South* (New Orleans, La.: Office of De Bow's Review, 1852), 308.
93. Lewis W. Hackett, *Malaria in Europe: An Ecological Study* (London: Oxford University Press, 1937), 175.
94. www.censusscope.org/us/map_common_race.html.
95. www.demographia.com/db-landstatepopdens.htm.

4. MALARIAL ECOLOGIES

1. Robert Sallares, *Malaria and Rome: A History of Malaria in Ancient Italy* (Oxford: Oxford University Press, 2002), 182.
2. Robert Sallares, "Role of Environmental Changes in the Spread of Malaria in Europe During the Holocene," *Quaternary International* 150 (2006): 21–27.

3. Richard Carter and Kamini Mendis, "Evolutionary and Historical Aspects of the Burden of Malaria," *Clinical Microbiology Reviews* 15, no. 4 (October 2002): 564–94.
4. Mario Coluzzi, "The Clay Feet of the Malaria Giant and Its African Roots," *Parassitologia* 41 (1999): 280.
5. Lewis W. Hackett, *Malaria in Europe: An Ecological Study* (London: Oxford University Press, 1937), 41.
6. Andrew Spielman and Michael D'Antonio, *Mosquito: A Natural History of Our Most Persistent and Deadly Foe* (New York: Hyperion, 2001), 5–6.
7. Ibid., 7–8, 41.
8. Competitive interactions between larvae of different *Anopheline* species that lead to increased mortality have been observed in laboratory settings. See C.J.M. Koenraadt et al., "The Effect of Food and Space on the Occurrence of Cannibalism and Predation Among Larvae of *Anopheles gambiae* sl," *Entomologia Experimentalis et Applicata* 112 (2004): 125–34.
9. Carter and Mendis, "Evolutionary and Historical Aspects of the Burden of Malaria," 564–94.
10. Leonard Jan Bruce-Chwatt and Julian de Zulueta, *The Rise and Fall of Malaria in Europe: A Historico-Epidemiological Study* (Oxford: Oxford University Press/ Regional Office for Europe of the World Health Organization, 1980), 89.
11. Sallares, *Malaria and Rome*, 97.
12. Ibid., 186.
13. Ibid., 134.
14. Robert Sallares, Abigail Bouwman, and Cecilia Anderung, "The Spread of Malaria to Southern Europe in Antiquity: New Approaches to Old Problems," *Medical History* 48 (2004): 311–28.
15. Sallares, *Malaria and Rome*, 134.
16. Alessandro Perosa et al., "Febris: A Poetic Myth Created by Poliziano," *Journal of the Warburg and Courtauld Institutes* 9 (1946): 86.
17. Spielman and D'Antonio, *Mosquito*, 49.
18. Sallares, *Malaria and Rome*, 4.
19. Interview with David Soren, January 11, 2007.
20. Bruce-Chwatt and de Zulueta, *The Rise and Fall of Malaria in Europe*, 13.
21. Sallares, *Malaria and Rome*, 103.
22. Coluzzi, "The Clay Feet of the Malaria Giant," 280.
23. Bruce-Chwatt and de Zulueta, *The Rise and Fall of Malaria in Europe*, 23, and R. Sallares, "Role of Environmental Changes in the Spread of Malaria in Europe During the Holocene," *Quaternary International* 150 (2006): 21–27.
24. Sallares, "Role of Environmental Changes in the Spread of Malaria in Europe," 21–27.

25. Ibid.
26. Sallares, *Malaria and Rome*, 49–53.
27. Ibid.
28. Perosa et al., "Febris," 88.
29. Ibid.; David Soren and Noelle Soren, eds., *A Roman Villa and a Late Roman Infant Cemetery: Excavation at Poggio Gramignano Lugnano in Teverina* (Rome: L'erma di Bretschneider, 1999), 648.
30. F. E. Romer, "Famine, Pestilence, and Brigandage in Italy in the Fifth Century AD," in Soren and Soren, eds., *A Roman Villa*, 465.
31. Ibid., 469.
32. Interview with David Soren, January 11, 2007.
33. David Soren, "Can Archaeologists Excavate Evidence of Malaria?" *World Archaeology* 35 (2003): 193–209.
34. Sallares, *Malaria and Rome*, 205.
35. Carter and Mendis, "Evolutionary and Historical Aspects of the Burden of Malaria," 564–94.
36. Paul Reiter, "From Shakespeare to Defoe: Malaria in England in the Little Ice Age," *Emerging Infectious Diseases* 6, no. 1 (January/February 2000): 1–14.
37. www.answers.com/topic/dante-alighieri.
38. Frank M. Snowden, *The Conquest of Malaria: Italy, 1900–1962* (New Haven, Conn.: Yale University Press, 2006), 39–40.
39. Fiammetta Rocco, *The Miraculous Fever-Tree: Malaria and the Quest for a Cure That Changed the World* (New York: HarperCollins, 2003), 36.
40. Ibid.
41. Sallares, *Malaria and Rome*, 53.
42. Ibid., 231.
43. Perosa et al., "Febris," 86.
44. Quoted in Sallares, *Malaria and Rome*, 227.
45. Ibid., 9, quoting *Letters of Horace Walpole*, ed. C. D. Yonge (1889).
46. Quoted in Daniel Pick, "'*Roma o Morte*': Garibaldi, Nationalism and the Problem of Psycho-biography," *History Workshop Journal* 57 (2004): 1–33.
47. Mary Keele, ed., *Florence Nightingale in Rome: Letters Written by Florence Nightingale in Rome in the Winter of 1847–1848* (Philadelphia, Penn.: American Philosophical Society, 1981), 27.
48. Ibid.
49. Quoted in Pick, "'*Roma o Morte*': Garibaldi, Nationalism and the Problem of Psycho-biography," 1–33.
50. Quoted in Sallares, *Malaria and Rome*, 176.
51. Snowden, *The Conquest of Malaria*, 33.
52. Ibid., 16.

53. Theodore Steinberg, *Nature Incorporated: Industrialization and the Waters of New England* (Cambridge, UK: Cambridge University Press, 1991), 12; and Walter H. Voskuil, *The Economics of Water Power Development* (Chicago and New York: A. W. Shaw Company, 1928), 3.
54. Ibid.
55. Interview with Susan McKnight, March 7, 2007; James Stevens Simmons, "The Transmission of Malaria by the *Anopheles* Mosquitoes of North America," in Forest Ray Moulton, ed., *A Symposium on Human Malaria with Special Reference to North America and the Caribbean Region* (Washington, D.C.: American Association for the Advancement of Science, 1941), 113–19; T.H.D. Griffitts, "Impounded Waters and Malaria," *Southern Medical Journal* 19 (1926): 367–70.
56. Todd L. Savitt and James Harvey Young, eds., *Disease and Distinctiveness in the American South* (Knoxville: University of Tennessee Press, 1988), 37.
57. John Frederick Schroeder, *Memoir of the Life and Character of Mrs. Mary Anna Boardman* (New Haven, Conn.: T. J. Stafford, 1849), 130.
58. Ibid., 130.
59. Rachel D. Carley, *Voices from the Past: A History as Told by the New Milford Historical Society's Portraits and Paintings* (West Kennebunk, Maine: New Milford Historical Society by Phoenix Pub., 2000), 44–48.
60. Ibid.
61. Ibid.
62. Ibid.
63. www.ftp.rootsweb.com/pub/usgenweb/ct/litchfield/history/1882/historyo/churchof44gms.txt.
64. See www.ftp.rootsweb.com/pub/usgenweb/ct/litchfield/history/1882/historyo/churchof44gms.txt; G. H. Waldrop, Jr., "Grist Mills of New Milford: Little Falls Mill," New Milford Historical Society, November 1998, and biographical data on Capt. Joseph Ruggles, New Milford Historical Society; Oliver Wendell Holmes, *Boylston Prize Dissertations for the Years 1836 and 1837* (Boston: Charles C. Little and James Brown, 1838), 55.
65. Holmes, *Boylston Prize Dissertations for the Years 1836 and 1837*, 56–57; see www.sots.ct.gov/RegisterManual/SectionVII/Population1756.htm.
66. Letter to Samuel Whiting, Esq., of Great Barrington, Mass., from Elijah Boardman, dated 1796. From New Milford Historical Society, Boardman Papers, folder 1.
67. Ibid.
68. David A. Warrell and Herbert M. Gilles, eds., *Essential Malariology*, 4th ed. (London: Hodder Arnold, 2002), 196.

69. John T. Cumbler, *Northeast and Midwest United States: An Environmental History* (Santa Barbara, Calif.: ABC-CLIO, Inc., 2005), 73.
70. Holmes, *Boylston Prize Dissertations for the Years 1836 and 1837*, 55.
71. Schroeder, *Memoir of the Life and Character of Mrs. Mary Anna Boardman*, 143.
72. Margaret Humphreys, *Malaria: Poverty, Race, and Public Health in the United States* (Baltimore: Johns Hopkins University Press, 2001), 37.
73. John Duffy, "Impact of Malaria on the South," in Savitt and Young, eds., *Disease and Distinctiveness in the American South*, 41.
74. Erwin H. Ackerknecht, *Malaria in the Upper Mississippi Valley, 1760–1900* (Baltimore: Johns Hopkins University Press, 1945), 56.
75. "Where Malaria Is Bred: The Underground Streams and Swamps of the City," *New York Times*, November 6, 1883.
76. "Malarial Sickness: The Fever in Long Island City," *New York Times*, October 1, 1877.
77. "Malaria's Baleful Work," *New York Times*, August 22, 1881.
78. "Where Malaria Is Bred."
79. Ibid.
80. "Long Island Malaria," *New York Times*, October 3, 1877.
81. "Where Malaria Is Bred."
82. "Malaria's Baleful Work."
83. Correspondence with Michael Raber, July 21, 2006.
84. Voskuil, *The Economics of Water Power Development*, 15.
85. Cumbler, *Northeast and Midwest United States*, 57.
86. Quoted in ibid.
87. Holmes, *Boylston Prize Dissertations for the Years 1836 and 1837*, 55.
88. W. V. King and G. H. Bradley, "Distribution of the Nearctic Species of Anopheles" and "Bionomics and Ecology of Nearctic Anopheles," in Forest Ray Moulton, ed., *A Symposium on Human Malaria with Special Reference to North America and the Caribbean Region* (Washington, D.C.: American Association for the Advancement of Science, 1941), 71–87.
89. Voskuil, *The Economics of Water Power Development*, 15.
90. "Malaria's Baleful Work."
91. C. M. Wenyon, "The Incidence and Etiology of Malaria in Macedonia," *Journal of the Royal Army Medical Corps* 27 (1921): 83–277.
92. Bruce-Chwatt and de Zulueta, *The Rise and Fall of Malaria in Europe*, 47.
93. H. Collinson Owen, *Salonika and After* (London: Hodder and Stoughton, 1919), 175–85.
94. Bruce-Chwatt and de Zulueta, *The Rise and Fall of Malaria in Europe*, 47.
95. Ibid.

96. Owen, *Salonika and After*, 175–85.
97. Robert Bwire, *Bugs in Armor: A Tale of Malaria and Soldiering* (Lincoln, Neb.: ToExcel Press, 2000), 38, and Bruce-Chwatt and de Zulueta, *The Rise and Fall of Malaria in Europe*, 47.
98. Bruce-Chwatt and de Zulueta, *The Rise and Fall of Malaria in Europe*, 47.
99. Owen, *Salonika and After*, 187.
100. Hackett, *Malaria in Europe*, 2.
101. Owen, *Salonika and After*, 186.
102. Wenyon, "The Incidence and Etiology of Malaria in Macedonia," 83–277.
103. Ackerknecht, *Malaria in the Upper Mississippi Valley, 1760–1900*, 89.
104. Wenyon, "The Incidence and Etiology of Malaria in Macedonia," 83–277.
105. Hackett, *Malaria in Europe*, 2.
106. Bwire, *Bugs in Armor*, 40.
107. A. B. Knudsen and R. Slooff, "Vector-borne Disease Problems in Rapid Urbanization: New Approaches to Vector Control," *Bulletin of the World Health Organization* 70, no. 1 (1992): 1–6.
108. Sir Malcolm Watson, *African Highway: The Battle for Health in Central Africa* (London: John Murray Publishers, 1953), 36.
109. Ibid., 24.
110. Ibid., 13–14.
111. Ibid., 26.
112. Steven Feierman, "Struggles for Control: The Social Roots of Health and Healing in Modern Africa," *African Studies Review* 28, no. 2/3 (June–September 1985): 119.
113. Watson, *African Highway*, 4, 174.
114. Pim Martens and Lisbeth Hall, "Malaria on the Move: Human Population Movement and Malaria Transmission," *Emerging Infectious Diseases* 6, no. 2 (March–April 2000): 103–109.
115. Amy Yomiko Vittor et al., "The Effect of Deforestation on the Human-Biting Rate of *Anopheles Darlingi*, the Primary Vector of *Falciparum* Malaria in the Peruvian Amazon," *American Journal of Tropical Medicine and Hygiene* 74, no. 1 (2005): 676–80.
116. Marcia Caldas de Castro et al., "Malaria Risk on the Amazon Frontier," *PNAS* 103, no. 7 (February 14, 2006): 2452–57; Wanderli P. Tadei et al., "Ecologic Observations on Anopheline Vectors of Malaria in the Brazilian Amazon," *American Journal of Tropical Medicine and Hygiene* 59, no. 2 (1998): 325–35.
117. Amy Yomiko Vittor et al., "The Effect of Deforestation on the Human-Biting Rate of *Anopheles Darlingi*, the Primary Vector of *Falciparum* Malaria in the Peruvian Amazon," 3–11.

118. Asnakew Kebede et al., "New Evidence of the Effects of Agro-ecologic Change on Malaria Transmission," *American Journal of Tropical Medicine and Hygiene* 73, no. 4 (2005): 676–80.
119. Deepa Suryanarayan, "Malaria Epidemic Set to Sting Mumbai," *Daily News and Analysis*, August 30, 2006.
120. Swatee Kher, "Malaria on the Rise, but No Outbreak," Express India online, July 5, 2006.
121. Indo-Asian News Service, "299 Malaria Deaths, 19 Dengue Deaths This Year in India," FreshNews.in, September 3, 2008; Sumitra Deb Roy, "Malaria Becoming Harder to Treat," *Daily News and Analysis* (India), August 24, 2008; Sumitra Deb Roy, "2 More Malaria Deaths in 24 Hrs," *Daily News and Analysis* (India), August 26, 2008.
122. Fred Pearce, "Science: It Bites, It Kills, It's Coming to Essex," *The Independent* (London), February 18, 2000.
123. "Climate Change Brings Back Malaria," ANSA.it, February 1, 2007.
124. Alastair McIndoe, "Malaria Goes Global as the World Gets Warmer," *The Straits Times* (Singapore), April 29, 2008.
125. See, for example, Paul Reiter et al., "Global Warming and Malaria: A Call for Accuracy," *The Lancet* 4 (June 2004): 323–24.
126. R. Sari Kovats et al., "El Niño and Health," *Lancet* 362, no. 9394 (November 1, 2003): 1481–89.
127. Ian Fisher, "Kisii Journal: Malaria, a Swamp Dweller, Finds a Hillier Home," *New York Times*, July 21, 1999.
128. Ibid.
129. Andrew K. Githenko and William Ndegwa, "Predicting Malaria Epidemics in the Kenyan Highlands Using Climate Data: A Tool for Decision Makers," *Global Change and Human Health* 2, no. 1 (2001): 54–63.
130. Ibid., 54–63.
131. Hong Chen et al., "New Records of *Anopheles arabiensis* Breeding on the Mount Kenya Highlands Indicate Indigenous Malaria Transmission," *Malaria Journal* 5 (March 7, 2006): 17; and Harold Ayodo, "Malaria Infections on the Rise," *The Standard*, October 5, 2006.
132. Joan H. Bryan et al., "Malaria Transmission and Climate Change in Australia," *Medical Journal of Australia* 164 (1996): 345–47; John Walker, "Malaria in a Changing World: An Australian Perspective," *International Journal of Parasitology* 28 (1998): 947–53.

5. PHARMACOLOGICAL FAILURE

1. Interview with John Thomas, BASF, December 6, 2005
2. David A. Warrell and Herbert M. Gilles, eds., *Essential Malariology*, 4th ed. (London: Hodder Arnold, 2002), 305–309.
3. Institute of Medicine, *Saving Lives, Buying Time: Economics of Malaria Drugs in an Age of Resistance* (Washington, D.C.: National Academies Press, 2004), 212.
4. "Herbicide Hope for Malaria," BBC News, January 31, 2003.
5. Gerald Tenywa, "Chimps Eat Herbs to Cure Malaria," AllAfrica.com, January 19, 2007; information on *Vernonia amygdalina* at www.fao.org.
6. Nina L. Etkin, "The Co-evolution of People, Plants, and Parasites: Biological and Cultural Adaptations to Malaria," *Proceedings of the Nutrition Society* 62 (2003): 311–17.
7. Fiammetta Rocco, *The Miraculous Fever-Tree: Malaria and the Quest for a Cure That Changed the World* (New York: HarperCollins, 2003), 77.
8. Alan Crozier et al., eds., *Plant Secondary Metabolites: Occurrence, Structure and Role in the Human Diet* (Oxford, UK: Blackwell Publishing, 2006), 102.
9. Warrell and Gilles, eds., *Essential Malariology*, 4th ed., 280–81.
10. Erwin H. Ackerknecht, *Malaria in the Upper Mississippi Valley, 1760–1900* (Baltimore: Johns Hopkins University Press, 1945), 107.
11. Quoted in Jon Kukla, "Kentish Agues and American Distempers: The Transmission of Malaria from England to Virginia in the Seventeenth Century," *Southern Studies* 25, no. 2 (Summer 1986): 135–47.
12. Frank M. Snowden, *The Conquest of Malaria: Italy, 1900–1962* (New Haven, Conn.: Yale University Press, 2006), 46.
13. Philip D. Curtin, *Disease and Empire: The Health of European Troops in the Conquest of Africa* (Cambridge, UK: Cambridge University Press, 1998), 58.
14. Quoted in Patrick Brantlinger, "Victorians and Africans: The Genealogy of the Myth of the Dark Continent," *Critical Inquiry* 12, no. 1 (Autumn 1985): 166–203.
15. Rocco, *The Miraculous Fever-Tree*, 77.
16. Roy Porter, *The Greatest Benefit to Mankind: A Medical History of Humanity* (New York: W. W. Norton, 1997), 230.
17. Paul Reiter, "From Shakespeare to Defoe: Malaria in England in the Little Ice Age," *Emerging Infectious Diseases* 6, no. 1 (January–February 2000): 1–11.
18. Mark Honigsbaum, *The Fever Trail: In Search of the Cure for Malaria* (Farrar, Straus and Giroux, 2001), 34; Reiter, "From Shakespeare to Defoe," 1–11.
19. Charles Morrow Wilson, "Quinine: Reborn in Our Hemisphere," *Harper's*, August 1943.

20. Rocco, *The Miraculous Fever-Tree*, 225–30.
21. Oliver Wendell Holmes, *Boylston Prize Dissertations for the Years 1836 and 1837* (Boston: Charles C. Little and James Brown, 1838), 30.
22. Letter from David Livingstone to Dr. James Ormiston McWilliam, November 28, 1860, published in *Transactions of the Epidemiological Society of London*, 1860, available at www.livingstoneonline.ucl.ac.uk.
23. Ibid.
24. Physicians in the United States recommended the antimalarial properties of coffee as late as 1884. "One wonders whether this supposed virtue of coffee was not instrumental in converting the Americans from tea to coffee drinking." Ackerknecht, *Malaria in the Upper Mississippi Valley, 1760–1900*, 123.
25. Honigsbaum, *The Fever Trail*, 57.
26. Ackerknecht, *Malaria in the Upper Mississippi Valley, 1760–1900*, 113. Converted into 2006 dollars using inflation calculator at www.westegg.com/inflation/.
27. Ibid., 120.
28. Rocco, *The Miraculous Fever-Tree*, 249.
29. Norman Taylor, *Cinchona in Java: The Story of Quinine* (New York: Greenberg, 1945), 50.
30. Ibid., 59–61.
31. Ibid., 38.
32. Ibid., 55.
33. Ibid., 51.
34. Ibid., 50, 62.
35. Ibid., 54.
36. Ibid., 61.
37. Ibid., 66.
38. Snowden, *The Conquest of Malaria*, 46–47.
39. M. L. Duran-Reynals, *The Fever Bark Tree* (New York: Doubleday and Co., 1946), 212–27.
40. Rocco, *The Miraculous Fever-Tree*, 110.
41. Ibid., 225–30.
42. Ibid., 249.
43. Taylor, *Cinchona in Java*, 75.
44. Duran-Reynals, *The Fever Bark Tree*, 212–27.
45. For example, when a Japanese-owned cinchona operation refused to join the cartel and sold quinine at a more affordable price—to the American Red Cross and others—the Kina Bureau attacked. It threatened to hinder the company's shipments of Java-grown bark to Tokyo, and ordered its producers to keep quinine off the market to drive the price back up. "Quinine Seized Here in

Anti-trust Drive," *New York Times*, March 24, 1928. In 1929, as a bounty of bark threatened to depress the price of quinine, the bureau ordered cinchona plantations to be destroyed. Again, between 1934 and 1937, it restricted the export of cinchona bark from Dutch-controlled Indonesia, and banned the export of planting material, lest anyone successfully culture cinchona elsewhere. Duran-Reynals, *The Fever Bark Tree*, 212–27.

46. "Hoover Warns World of Trade Wars," *New York Times*, January 10, 1926.
47. "New Move to End the Quinine Trust," *New York Times*, March 30, 1928, 16.
48. "Indictments out in Quinine Inquiry," *New York Times*, March 31, 1928, 21.
49. "Cinchona: Quinine to You," *Fortune*, January 25, 1934 (unsigned article but presumed to be authored by Norman Taylor, according to his "Biographical Note," at the New York Botanical Gardens Mertz Library).
50. Ibid.
51. Duran-Reynals, *The Fever Bark Tree*, 212–27.
52. "Cinchona: Quinine to You."
53. Ibid.
54. Patricia Barton, "Powders, Potions, and Tablets: The 'Quinine Fraud' in British India, 1890–1939," in James H. Mills and Patricia Barton, eds., *Drugs and Empires: Essays in Modern Imperialism and Intoxication, c. 1500–c. 1930* (New York: Palgrave Macmillan, 2007), 145.
55. Ibid., 156.
56. Ibid., 146.
57. Ibid., 145.
58. Sheila Zurbrigg, "Rethinking the Human Factor in Malaria Mortality: The Case of Punjab, 1868–1940," *Parassitologia* 36 (1994): 121–35.
59. Honigsbaum, *The Fever Trail*, 87.
60. Rocco, *The Miraculous Fever-Tree*, 107.
61. Ackerknecht, *Malaria in the Upper Mississippi Valley, 1760–1900*, 106–13.
62. Letter from David Livingstone to Dr. James Ormiston McWilliam, November 28, 1860, available at www.livingstoneonline.ucl.ac.uk.
63. www.en.wikipedia.org/wiki/Quinine#Dosing
64. David Livingstone and John Kirk, "Original Communications: Remarks on the African Fever on the River Zambezi," letter to the editor of the *Medical Times and Gazette*, November 12, 1859.
65. William Garden Blaikie, *The Personal Life of David Livingstone* (London: John Murray, 1880), digital version available on Project Gutenberg, at www.gutenberg.org/files/13262/13262-8.txt.
66. Curtin, *Disease and Empire*, 24, and Ackerknecht, *Malaria in the Upper Mississippi Valley, 1760–1900*, 107.
67. Warrell and Gilles, eds., *Essential Malariology*, 4th ed., 281.

68. Ibid., 198.
69. Mark Harrison, *Public Health in British India: Anglo-Indian Preventive Medicine, 1859–1914* (Cambridge, UK: Cambridge University Press, 1994), 163.
70. F. Bruneel, B. Gachot, M. Wolff, B. Régnier, M. Danis, F. Vachon, "Resurgence of Blackwater Fever in Long-term European Expatriates in Africa: Report of 21 Cases and Review," *Clinical Infectious Diseases* 32, no. 8 (April 15, 2001), 1133–40.
71. C. M. Wenyon, "The Incidence and Etiology of Malaria in Macedonia," *Journal of the Royal Army Medical Corps* 27 (1921): 83–277.
72. Quoted in Gordon Harrison, *Mosquitoes, Malaria and Man: A History of the Hostilities Since 1880* (New York: E. P. Dutton, 1978), 172.
73. Snowden, *The Conquest of Malaria*, 46.
74. Ibid., 73.
75. Ibid., 75.
76. Ibid., 74–75.
77. Quoted in Greer Williams, *The Plague Killers: Untold Stories of Three Great Campaigns Against Disease* (New York: Charles Scribner's Sons, 1969), 146.
78. Robert Aura Smith, "Trade Preference Sought by Leaders in Philippines," *New York Times*, September 23, 1934.
79. Wilson, "Quinine: Reborn in Our Hemisphere."
80. Rocco, *The Miraculous Fever-Tree*, 288.
81. Mark Harrison, "Medicine and the Culture of Command: The Case of Malaria Control in the British Army During the Two World Wars," *Medical History* 40 (1996): 437–52.
82. Duran-Reynals, *The Fever Bark Tree*, 232.
83. Harrison, "Medicine and the Culture of Command," 437–52.
84. Andrew Spielman and Michael D'Antonio, *Mosquito: A Natural History of Our Most Persistent and Deadly Foe* (New York: Hyperion, 2001), 142.
85. Wilson, "Quinine: Reborn in Our Hemisphere."
86. "Increase Is Seen in Malaria Fever," *New York Times*, April 11, 1942; "Malaria Hits 100,000,000," *New York Times*, October 18, 1942.
87. Raymond B. Fosdick, "Malaria Control," *The Scientific Monthly*, January 1944, 48; Harry Summers, "4 'Jalopy' Planes Last Bataan Hope," *New York Times*, April 22, 1942; Duran-Reynals, *The Fever Bark Tree*, 237–41.
88. Fosdick, "Malaria Control," 48.
89. Harrison, "Medicine and the Culture of Command," 437–52.
90. Fosdick, "Malaria Control," 48.
91. "The Quinine Cartel," *New York Times*, September 6, 1942, 6.
92. Duran-Reynals, *The Fever Bark Tree*, 232.
93. Williams, *The Plague Killers*, 145.

94. John Farley, *To Cast Out Disease: A History of the International Health Division of the Rockefeller Foundation (1913–1951)* (Oxford: Oxford University Press, 2004), 134.
95. Williams, *The Plague Killers*, 147.
96. Spielman and D'Antonio, *Mosquito*, 143.
97. www.sel.barc.usda.gov/diptera/ann_text.html.
98. Robert J. T. Joy, "Malaria in American Troops in the South and Southwest Pacific in World War II," *Medical History* 43 (1999): 192–207.
99. Institute of Medicine, *Saving Lives, Buying Time*, 260.
100. Laurie Garrett, *The Coming Plague: Newly Emerging Diseases in a World Out of Balance* (New York: Penguin Books, 1994), 49.
101. Robert S. Desowitz, *The Malaria Capers: More Tales of Parasites and People, Research and Reality* (New York: W. W. Norton, 1991), 205.
102. Letter from Norman Taylor to Cinchona Instituut, February 2, 1947, Norman Taylor Papers Archive, New York Botanical Garden, Mertz Library, Series 2.
103. C. P. Gilmore, "Malaria Wins Round 2," *New York Times*, September 25, 1966.
104. Institute of Medicine, *Saving Lives, Buying Time*, 173.
105. Desowitz, *The Malaria Capers*, 205.
106. Letter from Norman Taylor to Cinchona Instituut, January 11, 1946, Norman Taylor Papers Archive, New York Botanical Garden, Mertz Library, Series 2.
107. Letter from Norman Taylor to Cinchona Instituut, April 26, 1948, Norman Taylor Papers Archive, New York Botanical Garden, Mertz Library, Series 2.
108. Letter from Norman Taylor to Cinchona Instituut, August 10, 1948, Norman Taylor Papers Archive, New York Botanical Garden, Mertz Library, Series 2.
109. Letter from Norman Taylor to Cinchona Instituut, January 1, 1947, Norman Taylor Papers Archive, New York Botanical Garden, Mertz Library, Series 2.
110. A. W. Sweeney, "The Possibility of an 'X' Factor: The First Documented Drug Resistance of Human Malaria," *International Journal of Parasitology* 26, no. 10 (1996): 1035–61.
111. Ibid.
112. Jonathan D. Moreno, *Undue Risk: Secret State Experiments on Humans* (New York: W. H. Freeman and Company, 2000), 50.
113. Sweeney, "The Possibility of an 'X' Factor," 1035–61.
114. Ibid.
115. Ibid.
116. Ibid.
117. Ibid.
118. Ibid.
119. Ibid.
120. Ibid.

121. Ibid.
122. Walther H. Wernsdorfer, "Epidemiology of Drug Resistance in Malaria," *Acta Tropica* 56 (1994): 143–56.
123. Sweeney, "The Possibility of an 'X' Factor," 1035–61.
124. Elisabeth Rosenthal, "Outwitted by Malaria, Desperate Doctors Seek New Remedies," *New York Times*, February 12, 1991.
125. Gilmore, "Malaria Wins Round 2."
126. Institute of Medicine, *Saving Lives, Buying Time*, 260.
127. Elisabeth Rosenthal, "Outwitted by Malaria."
128. I. Singh and T. S. Kalyanum, "The Superiority of 'Camoquin' over Other Antimalarials," *British Medical Journal* 2, no. 4779 (August 9, 1952): 312–15.
129. C. M. Trenholme et al., "Mefloquine (WR 142,490) in the Treatment of Human Malaria," *Science* 190, no. 4216 (November 21, 1975): 792–94.
130. T. M. Cosgriff et al., "Evaluation of the Antimalarial Activity of the Phenanthrenemethanol Halofantrine (WR 171,669)," *American Journal of Tropical Medicine and Hygiene* 31, no. 6 (November 1982): 1075–79.
131. N. J. White, "Quinidine in Falciparum Malaria," *Lancet* 318, no. 8255 (November 1981): 1069–71.
132. Cosgriff, "Evaluation of the Antimalarial Activity of the Phenanthrenemethanol Halofantrine (WR 171,669)," 1075–97.
133. U. D'Alessandro and H. Buttiëns, "History and Importance of Antimalarial Drug Resistance," *Tropical Medicine and International Health* 6, no. 11 (November 2001): 845–48.
134. Rosenthal, "Outwitted by Malaria."
135. Wernsdorfer, "Epidemiology of Drug Resistance in Malaria," 143–56.
136. Pamela McElwee, "'There Is Nothing That Is Difficult': History and Hardship on and After the Ho Chi Minh Trail in North Vietnam," *Asia Pacific Journal of Anthropology* 6, no. 3 (December 2005): 197–214.
137. Ibid.
138. "Military Scientist Took War on Malaria from Jungle to Market," *South China Morning Post*, January 1, 2006, 5.
139. John Prados, *The Blood Road: The Ho Chi Minh Trail and the Vietnam War* (New York: John Wiley and Sons, 1999), xiv.
140. D'Alessandro and Buttiëns, "History and Importance of Antimalarial Drug Resistance," 845–48, and Walter Modell, "Malaria and Victory in Vietnam," *Science* 162, no. 3860 (December 1968): 1346–52; Gilmore, "Malaria Wins Round 2."
141. Vivien Cui, "Military Scientist Took War on Malaria from Jungle to Market," *South China Morning Post*, January 1, 2006, 5; Merrill Goozner, "The First 13-Year-Old Patient," *The Scientist* 20, no. 12 (December 2006).

142. Qinghaosu Antimalaria Coordinating Research Group, "Antimalaria Studies on Qinghaosu," *Chinese Medical Journal* 92 (December 1979): 811–16.
143. Ibid.
144. Elisabeth Hsu, "Reflections on the 'Discovery' of the Antimalarial *Qinghao*," *British Journal of Clinical Pharmacology* 61, no. 6 (June 2006): 666–70.
145. T. T. Hien and N. J. White, "Qinghaosu," *The Lancet* 341, no. 8845 (March 6, 1993): 603–608.
146. Cui, "Military Scientist Took War on Malaria from Jungle to Market"; Qinghaosu Antimalaria Coordinating Research Group, "Antimalaria Studies on Qinghaosu."
147. Hsu, "Reflections on the 'Discovery' of the Antimalarial *Qinghao*," 666–70.
148. Ibid.
149. Qinghaosu Antimalaria Coordinating Research Group, "Antimalaria Studies on Qinghaosu."
150. Walther H. Wernsdorfer, "Drug Resistance of Malaria Parasites," Twentieth Expert Committee on Malaria, working paper MAL/ECM/20/98/15, undated.
151. Médecins Sans Frontières, "ACT NOW to Get Malaria Treatment That Works to Africa," April 2003.
152. "Herbal Vietnam War Remedy Key to Cheap Malaria Cure," *Edmonton Journal*, November 16, 2003.
153. David Lague, "Revolutionary Discovery," *Far Eastern Economic Review*, March 14, 2002.
154. Goozner, "The First 13-Year-Old Patient."
155. Qinghaosu Antimalaria Coordinating Research Group, "Antimalaria Studies on Qinghaosu."
156. Hsu, "Reflections on the 'Discovery' of the Antimalarial *Qinghao*," 666–70.
157. Lague, "Revolutionary Discovery."
158. Carrie Chan, "Malaria Expert Close to Achieving His Dream," *South China Morning Post*, March 6, 2003.
159. Andrew Jack, "Monotherapy 'Saves the Lives of So Many,'" *Financial Times*, January 20, 2006, 10.
160. Cui, "Military Scientist Took War on Malaria from Jungle to Market," 5.
161. Ibid.
162. "Novartis Malaria Drug Riamet Wins Marketing Approval in Switzerland"; "Novartis in Talks with WHO over Cheaper Malaria Drugs"; Agence France Presse, May 3, 2001.
163. "Novartis Malaria Drug Riamet Wins Marketing Approval in Switzerland"; "Novartis in Talks with WHO over Cheaper Malaria Drugs"; Gatonye Gathura, "WHO Warns Malaria Drug Makers," *The Nation* (Kenya), January 26, 2006; Jack, "Monotherapy 'Saves the Lives of So Many,'" 10.

164. Institute of Medicine, *Saving Lives, Buying Time*, 174
165. Médecins Sans Frontières, "ACT NOW to Get Malaria Treatment That Works to Africa."
166. Institute of Medicine, *Saving Lives, Buying Time*, 175.
167. Amir Attaran et al., "WHO, the Global Fund, and Medical Malpractice in Malaria Treatment," *Lancet* 363 (January 17, 2004): 237–40.
168. Médecins Sans Frontières, "ACT NOW to Get Malaria Treatment That Works to Africa."
169. Ibid.
170. Melody Peterson, "Novartis Agrees to Lower Price of a Medicine Used in Africa," *New York Times*, May 3, 2001.
171. Donald G. McNeil, "New Drug for Malaria Pits U.S. Against Africa," *New York Times*, May 28, 2002.
172. Rick Steketee, "Policy Change to Use Effective Antimalarial Drugs in Programs—CDC Experience," Roll Back Malaria partners meeting, Geneva, February 26–28, 2002, available at www.rbm.who.int/docs/5pm_presentations/Steketee.ppt.
173. Donald G. McNeil, "Herbal Drug Widely Embraced in Treating Resistant Malaria," *New York Times*, May 10, 2004, 1.
174. Gavin Yamey, "Global Health Agencies Are Accused of Incompetence," *British Medical Journal* 321, no. 7264 (September 30, 2000): 787.
175. Attaran et al., "WHO, the Global Fund, and Medical Malpractice in Malaria Treatment."
176. Institute of Medicine, *Saving Lives, Buying Time*, 313.
177. "WHO Calls for an Immediate Halt to Provision of Single-drug Artemisinin Malaria Pills," press release, January 20, 2006.
178. McNeil, "New Drug for Malaria Pits U.S. Against Africa."
179. Yamey, "Global Health Agencies Are Accused of Incompetence."
180. Jeanne Whalen, "Novartis Cuts Price of Coartem to Help Fight Malaria in Africa," *Wall Street Journal*, October 2, 2006.
181. Donald G. McNeil, "Drug Partnership Introduces Cheap Antimalaria Pill," *New York Times*, March 1, 2007.
182. Institute of Medicine, *Saving Lives, Buying Time*, 9.
183. Global Fund to Fight AIDS, Tuberculosis and Malaria, Global Fund Eighteenth Board Meeting Decision Point, GF/B18/DP7, Eighteenth Board Meeting, New Delhi, India, November 7–8, 2008.
184. Institute of Medicine, *Saving Lives, Buying Time*, 116.
185. Médecins Sans Frontières, "ACT NOW to Get Malaria Treatment That Works to Africa."

186. Gathura, "WHO Warns Malaria Drug Makers."
187. Jack, "Monotherapy 'Saves the Lives of So Many,'" *Financial Times*, January 20, 2006, 10.
188. Ibid.
189. Institute of Medicine, *Saving Lives, Buying Time*, 317.
190. Médecins Sans Frontières, "ACT NOW to Get Malaria Treatment That Works to Africa."
191. McNeil, "Herbal Drug Widely Embraced in Treating Resistant Malaria."
192. Ferrer-Rodriguez et al., "*Plasmodium Yoelii*: Identification and Partial Characterization of an *MDR1* Gene in an Artemisinin-Resistant Line," *Journal of Parasitology* 90 (2004): 152–60.
193. Patrick E. Duffy and Carol Hopkins Sibley, "Are We Losing Artemisinin Combination Therapy Already?" *Lancet* 366, no. 9501 (December 3, 2005): 1908–909; Ronan Jambou et al., "Resistance of *Plasmodium falciparum* Field Isolates to in-vitro Artemether and Point Mutations of the SERCA-type PfATPase6," *Lancet* 366, no. 9501 (December 3, 2005): 1960–63.
194. Duffy and Sibley, "Are We Losing Artemisinin Combination Therapy Already?" 1908–09; Jambou et al., "Resistance of *Plasmodium falciparum* Field Isolates," 1960–63.
195. World Health Organization, Report of an Informal Consultation, "Containment of Malaria Multi-drug Resistance on the Cambodia-Thailand Border," Phnom Penh, January 29–30, 2007. In January 2009, a study reported that ACT-resistant *falciparum* had spread beyond the Cambodia-Thai border and deeper into southern Cambodia. W. O. Rogers et al., "Failure of Artesunate-mefloquine Combination Therapy for Uncomplicated *Plasmodium falciparum* Malaria in Southern Cambodia," *Malaria Journal* 8, no. 10 (January 12, 2009).
196. Donald G. McNeil, "Drug Makers Get a Warning from the U.N. Malaria Chief," *New York Times*, January 20, 2006.
197. Gathura, "WHO Warns Malaria Drug Makers."
198. Nicholas Zamiska, "Infectious Issue: Global Health, China's Pride on Line in Malaria Clash," *Wall Street Journal*, March 6, 2007.
199. Donald G. McNeil, "An Iron Fist Joins the Malaria Wars," *New York Times*, June 27, 2006.
200. Paul N. Newton et al., "Manslaughter by Fake Artesunate in Asia—Will Africa Be Next?" *PLoS Medicine* 3, no. 6 (June 2006).
201. Paul N. Newton et al., "A Collaborative Epidemiological Investigation into the Criminal Fake Artesunate Trade in South East Asia," *PLoS Medicine* 5, no. 2 (February 2008); Walt Bogdanich and Jake Hooker, "Battle Against Counterfeit Drugs Has New Weapon: Pollen," *New York Times*, February 12,

2008; Reuters, "Fake Malaria Drugs Threatening Africa, Says Expert," Gulfnews.com, July 24, 2006; Newton et al., "Manslaughter by Fake Artesunate in Asia—Will Africa Be Next?"

202. Willard H. Wright, *Forty Years of Tropical Medicine Research: A History of the Gorgas Memorial Institute of Tropical and Preventive Medicine, Inc., and the Gorgas Memorial Laboratory* (Baltimore, Md.: Reese Press, 1970), 52.
203. Richard M. Garfield and Sten H. Vermund, "Changes in Malaria Incidence After Mass Drug Administration in Nicaragua," *Lancet*, 2, no. 8348 (1983): 500–03.
204. David Lague, "On Island Off Africa, China Hopes to Wipe Out Malaria," *International Herald Tribune*, June 6, 2007.
205. Ed Harris, "Chinese Researchers Claim Comoros Malaria Success," Reuters, March 11, 2008.

6. THE KARMA OF MALARIA

1. David Ropeik, "Understanding Factors of Risk Perception," *Nieman Reports*, Winter 2002.
2. Author visit to Chikwawa, Malawi, February 23, 2007.
3. John Peffer-Engels, *Chewa* (New York: Rosen Publishing Group, 1996), 17–18.
4. Deborah Kaspin, "A Chewa Cosmology of the Body," *American Ethnologist* 23, no. 3 (August 1996): 561–78.
5. Interview with Dr. Yamikani Chimalizeni, Blantyre, Malawi, February 19, 2007.
6. Deborah L. Helitzer et al., "The Role of Ethnographic Research in Malaria Control: An Example from Malawi," *Research in the Sociology of Health Care*, 10 (1993): 269–86.
7. H. Kristian Heggenhougen et al, *The Behavioural and Social Aspects of Malaria and Its Control* (Geneva, Switzerland: World Health Organization, 2003), 43.
8. Interview with David Smith, Blantyre, Malawi, February 24, 2007.
9. Steven Feierman, "Struggles for Control: The Social Roots of Health and Healing in Modern Africa," *African Studies Review* 28, no. 2/3 (June–September 1985): 87.
10. Peffer-Engels, *Chewa*, 19.
11. Helitzer et al., "The Role of Ethnographic Research in Malaria Control," 269–86.
12. Institute of Medicine, *Saving Lives, Buying Time: Economics of Malaria Drugs in an Age of Resistance* (Washington, D.C.: National Academies Press, 2004), 58.

13. Heggenhougen et al., *The Behavioural and Social Aspects of Malaria and Its Control,* 10.
14. June Msechu, "Community's Perceptions and Use of Antimalarial Drugs in the Home Management of Malaria in Rural Tanzania," Fourth MIM Pan-African Malaria Conference, Yaoundé, Cameroon, November 15, 2005.
15. Institute of Medicine, *Saving Lives, Buying Time,* 7.
16. Heggenhougen et al., *The Behavioural and Social Aspects of Malaria and Its Control,* 50, 56.
17. Helitzer et al., "The Role of Ethnographic Research in Malaria Control," 269–86.
18. Ibid.
19. M. Ettling et al., "Economic Impact of Malaria in Malawian Households," *Tropical Medicine and Parasitology* 45 (1994): 74–79.
20. Ibid.
21. Heggenhougen et al., *The Behavioural and Social Aspects of Malaria and Its Control,* 136.
22. Michael and Elspeth King, *The Story of Medicine and Disease in Malawi: The 130 Years Since Livingstone* (Blantyre, Malawi: Montfort Press, 1992), 23–25.
23. Author visit to Chikwawa, Malawi, February 23, 2007.
24. Msechu, "Community's Perceptions and Use of Antimalarial Drugs."
25. Tidiane Ndoye, "L'observance des traitements antipaludiques au Senegal: Le rôle des differents dispensateurs de traitements," Fourth MIM Pan-African Malaria Conference, Yaoundé, Cameroon, November 15, 2005.
26. Interview with Dr. Yamikani Chimalizeni, Blantyre, Malawi, February 19, 2007.
27. Heggenhougen et al., *The Behavioural and Social Aspects of Malaria and Its Control,* 60.
28. Caroline Jones, "The Social Reality of Malaria," Fourth MIM Pan-African Malaria Conference, Yaoundé, Cameroon, November 15, 2005.
29. Author visit to Gorgas Memorial Institute, Panama City, Panama, April 21, 2006.
30. Sir Malcolm Watson, *African Highway: The Battle for Health in Central Africa* (London: John Murray Publishers, 1953), 236.
31. Presentation by Kamija Phiri, Queen Elizabeth Hospital, Blantyre, Malawi, February 13, 2007.
32. Interview with Dr. Themba Mzilahowa, Blantyre, Malawi, February 14, 2007.
33. World Health Organization, *World Malaria Report 2008,* Geneva,10.
34. Institute of Medicine, *Saving Lives, Buying Time,* 221.
35. Lars Hviid, "Immunology and Pathogenesis: Naturally Acquired Protective Immunity to *Plasmodium falciparum* Malaria in Africa," Fourth MIM Pan-African Malaria Conference, Yaoundé, Cameroon, November 16, 2005.

36. Institute of Medicine, *Saving Lives, Buying Time*, 146.
37. Interview with Terrie Taylor, Blantyre, Malawi, February 2007.
38. Institute of Medicine, *Saving Lives, Buying Time*, 221.
39. Ibid., 169.
40. Donald McNeil, "Revisions Sharply Cut Estimates on Malaria," *New York Times*, September 23, 2008.
41. Interview with Socrates Litsios, New Haven, Conn., November 8, 2008.
42. Jones, "The Social Reality of Malaria."
43. Sheldon Watts, *Epidemics and History: Disease, Power, and Imperialism* (New Haven, Conn.: Yale University Press, 1997), 225.
44. Donald G. McNeil, Jr., "Drug Partnership Introduces Cheap Antimalaria Pill," *New York Times*, March 1, 2007, A3. Standing in front of a line of Malawian women and children melting onto a hospital bench waiting to hear results of their malaria tests, the malariologist Karl Seydel remarked, "They'd be relieved to get a positive." Interview with Karl Seydel, February 2007.
45. Interview with Martin Hayman, November 15, 2006.
46. Institute of Medicine, *Saving Lives, Buying Time*, 314.
47. Gerard Krause and Rainer Sauerborn, "Comprehensive Community Effectiveness of Health Care: A Study of Malaria Treatment in Children and Adults in Rural Burkina Faso," *Annals of Tropical Paediatrics* 20 (2000): 273–82.
48. Heggenhougen et al., *The Behavioural and Social Aspects of Malaria and Its Control*, 151.
49. W. Fungladda, "Health Behaviour and Illness Behaviour of Malaria: A Review," quoted in Heggenhougen et al., *The Behavioural and Social Aspects of Malaria and Its Control*, 11.
50. Tina Rosenberg, "The Scandal of 'Poor People's Diseases,'" *New York Times*, March 29, 2006.
51. Patrick Brantlinger, "Victorians and Africans: The Genealogy of the Myth of the Dark Continent," *Critical Inquiry* 12, no. 1 (Autumn 1985): 166–203.
52. Michael and Elspeth King, *The Story of Medicine and Disease in Malawi*, quoting Robert Laws, 40.
53. David Livingstone and John Kirk, "Original Communications: Remarks on the African Fever on the River Zambezi," letter to the editor of *The Medical Times and Gazette*, November 12, 1859.
54. "Dr. Livingstone, the Great Explorer of Central Africa," *Harper's Weekly*, January 31, 1857.
55. Letter from David Livingstone to Dr. James Ormiston McWilliam, November 28, 1860, available at www.livingstoneonline.ucl.ac.uk, published in *Transactions of the Epidemiological Society of London*, 1860.

56. Michael and Elspeth King, *The Story of Medicine and Disease in Malawi*, 5, quoting Livingstone.
57. Quoted in R. M. Packard, "Malaria Dreams: Postwar Visions of Health and Development in the Third World," *Medical Anthropology* 17 (1997): 179–96.
58. Herbert S. Klein, *The Atlantic Slave Trade* (Cambridge, UK: Cambridge University Press, 1999), 185.
59. Michael and Elspeth King, *The Story of Medicine and Disease in Malawi*, 30.
60. Jeanne Whalen, "Novartis Cuts Price of Coartem to Help Fight Malaria in Africa," *Wall Street Journal*, October 2, 2006.
61. Interview with Bob Laverty, Yaoundé, Cameroon, November 12, 2005.
62. "Developing Countries Slow to Order Coartem Despite Boost in Production," Kaisernetwork.org, January 19, 2006.
63. Andrew Jack, "Up to 10m Malaria Tablets 'May Be Destroyed,'" *Financial Times*, July 24, 2006.

7. SCIENTIFIC SOLUTIONS

1. Visit to Dyann Wirth lab, Harvard School of Public Health, Cambridge, Mass., March 10, 2008.
2. Roy Porter, *The Greatest Benefit to Mankind: A Medical History of Humanity* (New York: W. W. Norton, 1997), 437.
3. Daniel Pick, "'*Roma o morte*': Garibaldi, Nationalism and the Problem of Psycho-biography," *History Workshop Journal* 57 (2004): 1–33.
4. Frank M. Snowden, *The Conquest of Malaria: Italy, 1900–1962* (New Haven, Conn.: Yale University Press, 2006), 21.
5. Ibid., 39–40.
6. Dale C. Smith and Lorraine B. Sanford, "Laveran's Germ: The Reception and Use of a Medical Discovery," *American Journal of Tropical Medicine and Hygiene* 34, no. 1 (1985): 2–20.
7. "Introduction: Part One, Recent Research in Malaria," *British Medical Bulletin* 8, no. 1 (1951); Robert S. Desowitz, *The Malaria Capers: More Tales of Parasites and People, Research and Reality* (New York: W. W. Norton, 1991), 167–68, and Gordon Harrison, *Mosquitoes, Malaria and Man: A History of the Hostilities Since 1880* (New York: E. P. Dutton, 1978), 11.
8. Douglas M. Haynes, *Imperial Medicine: Patrick Manson and the Conquest of Tropical Disease* (Philadelphia: University of Pennsylvania Press, 2001), 14.
9. Smith and Sanford, "Laveran's Germ," 2–20.
10. "North River Malaria," *New York Times*, July 13, 1884.
11. Smith and Sanford, "Laveran's Germ," 2–20.

12. Desowitz, *The Malaria Capers*, 169.
13. Haynes, *Imperial Medicine*, 14, 98.
14. Ibid., 22–40.
15. Ibid., 48, 51.
16. David Soren and Noelle Soren, eds., *A Roman Villa and a Late Roman Infant Cemetery: Excavation at Poggio Gramignano Lugnano in Teverina* (Rome: L'erma di Bretschneider, 1999), 637.
17. Stephen H. Gillespie and Richard D. Pearson, *Principles and Practice of Clinical Parasitology* (Hoboken, N.J.: John Wiley and Sons, 2001), 9.
18. John Farley, "Parasites and the Germ Theory of Disease," *Milbank Quarterly* 67, suppl. 1 (1989): 50–68.
19. Haynes, *Imperial Medicine*, 49–82.
20. James Stevens Simmons, *Malaria in Panama* (Baltimore: Johns Hopkins University Press, 1939), 8–9.
21. Ibid., 18; David McCullough, *The Path Between the Seas: The Creation of the Panama Canal, 1870–1914* (New York: Simon & Schuster, 1977), 134.
22. McCullough, *The Path Between the Seas*, 159, 174.
23. Press dispatch from Panama, "Is M. de Lesseps a Canal Digger or a Grave Digger?" *Harper's Weekly*, September 3, 1881.
24. McCullough, *The Path Between the Seas*, 144.
25. "North River Malaria."
26. See, e.g., R. Patterson, "Dr. William Gorgas and His War with the Mosquito," *Canadian Medical Association Journal* 141 (6), September 15, 1989, 596.
27. McCullough, *The Path Between the Seas*, 221–23.
28. Patrick Manson, "On the Nature and Significance of the Crescentic and Flagellated Bodies in Malarial Blood," *British Medical Journal* (December 8, 1894).
29. Haynes, *Imperial Medicine*, 90.
30. W. F. Bynum and Caroline Overy, eds., *The Beast in the Mosquito: The Correspondence of Ronald Ross and Patrick Manson* (Amsterdam and Atlanta, Calif.: Rodopi B.V., 1998), 55, 125.
31. Mark Harrison, *Public Health in British India: Anglo-Indian Preventive Medicine, 1859–1914* (London: Wellcome Institute for the History of Medicine), 1994.
32. Ronald Ross, *Child of Ocean: A Romance* (London: George Allen and Unwin Ltd., 1932), 118.
33. Bynum and Overy, eds., *The Beast in the Mosquito*, 321.
34. Ronald Ross, "The Third Element of the Blood and the Malaria Parasite," *Indian Medical Gazette*, January 1894, 5–14.
35. Bynum and Overy, eds., *The Beast in the Mosquito*, 5.
36. Ibid., 1–321.

37. Amico Bignami, "Hypotheses as to the Life-history of the Malarial Parasite Outside the Human Body," *Lancet* (November 21, 1896).
38. Ernesto Capanna, "Grassi *versus* Ross: Who Solved the Riddle of Malaria?" *International Microbiology* 9 (2006): 69–74; Ronald Ross, *Memoirs* (London: John Murray, 1928), 337.
39. Bignami, "Hypotheses as to the Life-history of the Malarial Parasite."
40. Bynum and Overy, eds., *The Beast in the Mosquito*, 85.
41. Ibid., 133.
42. Ibid., 279.
43. Ibid., 291.
44. Ibid., 336.
45. Ross, *Memoirs*, 339.
46. Bynum and Overy, eds., *The Beast in the Mosquito*, 355.
47. Ibid., 387.
48. Ibid., 341.
49. Bynum and Overy, eds., *The Beast in the Mosquito*, 357.
50. Ross, *Memoirs*, 341.
51. Bynum and Overy, eds., *The Beast in the Mosquito*, 411.
52. Ross, *Memoirs*, 355.
53. Bynum and Overy, eds., *The Beast in the Mosquito*.
54. Ross, *Memoirs*, 455.
55. Bynum and Overy, eds., *The Beast in the Mosquito*, 396.
56. Ibid., 289.
57. Capanna, "Grassi *versus* Ross," 69–74.
58. Bynum and Overy, eds., *The Beast in the Mosquito*, 466.
59. Lewis W. Hackett, *Malaria in Europe: An Ecological Study* (London: Oxford University Press, 1937), 12.
60. Ibid., 12.
61. Bynum and Overy, eds., *The Beast in the Mosquito*, 437.
62. J. M. Hurley, "Is the Mosquito a Disseminator of Malaria?" *Pacific Medical Journal* 48 (1905): 338–42, quoted in F. Ellis McKenzie and Ebrahim M. Samba, "The Role of Mathematical Modeling in Evidence-based Malaria Control," *American Journal of Tropical Medicine and Hygiene* 71, suppl. 2 (2004): 94–96.
63. Bynum and Overy, eds., *The Beast in the Mosquito*, 390.
64. Ross, *Memoirs*, 365.
65. Harrison, *Public Health in British India*, 159.
66. Ross, *Memoirs*, 430.
67. Bynum and Overy, eds., *The Beast in the Mosquito*, 449.
68. Quoted in N. H. Swellengrebel, "How the Malaria Service in Indonesia Came into Being, 1898–1948," *The Journal of Hygiene* 48, no. 2 (June 1950): 146–57.

69. Snowden, *The Conquest of Malaria*, 46.
70. Hackett, *Malaria in Europe*, 16.
71. Ross, *Memoirs*, 415.
72. S. R. Christophers, Second Report of the Anti-malarial Operations at Mian Mir, 1901–1903: Scientific Memoirs by the Officers of the Medical and Sanitary Departments of the Government of India, 1904.
73. Bynum and Overy, eds., *The Beast in the Mosquito*; W. F. Bynum, "'Reasons for Contentment': Malaria in India, 1900–1920," *Parassitologia* 40 (1998): 19–27.
74. Snowden, *The Conquest of Malaria*, 158.
75. Sir Malcolm Watson, *African Highway: The Battle for Health in Central Africa* (London: John Murray Publishers, 1953), 236.
76. James Whorton, *Before Silent Spring: Pesticides and Public Health in Pre-DDT America* (Princeton, N.J.: Princeton University Press, 1974), 6–41.
77. L. O. Howard, *Fighting the Insects: The Story of an Entomologist* (New York: Arno Press, 1980).
78. Andrew Spielman and Michael D'Antonio, *Mosquito: A Natural History of Our Most Persistent and Deadly Foe* (New York: Hyperion, 2001), 118; Gordon Harrison, *Mosquitoes, Malaria and Man*, 168.
79. John Farley, *To Cast Out Disease: A History of the International Health Division of the Rockefeller Foundation (1913–1951)* (Oxford: Oxford University Press, 2004), 111.
80. Ibid., 118.
81. Quoted in Hugh Evans, "European Malaria Policy in the 1920s and 1930s: The Epidemiology of Minutiae," *Isis* 80, no. 301 (March 1989): 40–59.
82. Gordon Harrison, *Mosquitoes, Malaria and Man*, 185.
83. Hackett, *Malaria in Europe*, 16.
84. Howard, *Fighting the Insects*, 120.
85. Hackett, *Malaria in Europe*, 40–41.
86. Paul F. Russell, "Identification of the Larvae of the Three Common *Anopheline* Mosquitoes of the Southern United States," *American Journal of Epidemiology* 5 (March 1925): 149–74.
87. Hackett, *Malaria in Europe*, 273.
88. Ibid., 235.
89. Ibid., 266.
90. "Will Tropical Medicine Move to the Tropics?" *Lancet* 347, no. 9002 (1995).
91. David Arnold, *The New Cambridge History of India: Science, Technology, and Medicine in Colonial India* (Cambridge, UK: Cambridge University Press, 2000), 145–46.
92. E. Richard Brown, "Public Health in Imperialism: Early Rockefeller Programs

at Home and Abroad," *American Journal of Public Health* 66, no. 9 (1976): 897–903.

93. William E. Collins and John W. Barnwell, "A Hopeful Beginning for Malaria Vaccines," *New England Journal of Medicine* 359 (December 11, 2008): 2599–601; Judith E. Epstein, "What Will a Partly Protective Malaria Vaccine Mean to Mothers in Africa?" *Lancet* 370, no. 9598 (November 3, 2007): 1523–24.
94. Ben C. L. van Schaijk et al., "Gene Disruption of *Plasmodium falciparum p52* Results in Attenuation of Malaria Liver Stage Development in Cultured Primary Human Hepatocytes," *PLoS One* 3, no. 10 (October 28, 2008); Jason Fagone, "The Scientist Ending Malaria with His Army of Mosquitoes," *Esquire*, December 8, 2008.
95. Louis Miller, quoted in Susan Okie, "Betting on a Malaria Vaccine," *New England Journal of Medicine* 353 (November 3, 2005): 1877–81.
96. Mary Moran et al., *The Malaria Product Pipeline: Planning for the Future* (Sydney, Australia: George Institute for International Health, 2007).
97. Interview with Dyann Wirth, March 10, 2008.
98. Kathryn S. Aultman et al., "*Anopheles gambiae* Genome: Completing the Malaria Triad," *Science* 298, no. 5591 (October 4, 2002): 13; "Gateses Give Record $5 Billion Gift to Foundation," *New York Times*, June 3, 1999.
99. Visit to Dyann Wirth lab, Harvard School of Public Health, Cambridge, Mass., March 10, 2008.

8. THE DISAPPEARED: HOW MALARIA VANISHED FROM THE WEST

1. M. J. Dobson, "'Marsh Fever': The Geography of Malaria in England," *Journal of Historical Geography*, 6, no. 4 (1980): 357–89.
2. Paul Reiter, "From Shakespeare to Defoe: Malaria in England in the Little Ice Age," *Emerging Infectious Diseases* 6, no. 1 (January–February 2000): 1–11.
3. Dobson, "'Marsh Fever,'" 357–89.
4. Ibid., 357–89.
5. Lewis W. Hackett, *Malaria in Europe: An Ecological Study* (London: Oxford University Press, 1937), 174.
6. Jon Kukla, "Kentish Agues and American Distempers: The Transmission of Malaria from England to Virginia in the Seventeenth Century," *Southern Studies* 25, no. 2 (Summer 1986): 135–47.
7. William MacArthur, "A Brief Story of English Malaria," *British Medical Bulletin* 8, no. 1 (1951): 76–79.

8. Stephen Frenkel and John Western, "Pretext or Prophylaxis? Racial Segregation and Malarial Mosquitoes in a British Tropical Colony: Sierra Leone," *Annals of the Association of American Geographers* 78, no. 2 (June 1988): 211–28; J.W.W. Stephens and S. R. Christophers, "The Segregation of Europeans," Report to the Malaria Committee of the Royal Society, October 1, 1900.
9. John W. Cell, "Anglo-Indian Medical Theory and the Origins of Segregation in West Africa," *The American Historical Review* 91, no. 2 (April 1986): 307–35.
10. Mark Harrison, "Medicine and the Culture of Command: The Case of Malaria Control in the British Army During the Two World Wars," *Medical History* 40 (1996): 437–52.
11. W. E. Baker, T. E. Dempster, and H. Yule, "The Prevalence of Organic Disease of the Spleen as a Test for Detecting Malarious Localities in Hot Climates" (Calcutta: Office of Superintendent of Government Printing, 1868), 14; Ian Stone, *Canal Irrigation in British India* (Cambridge, UK: Cambridge University Press, 1984), 18–20.
12. Elizabeth Whitcombe, *Agrarian Conditions in Northern India: The United Provinces Under British Rule, 1860–1900* (Berkeley: University of California Press, 1972), 62–64, 88.
13. David Arnold, *The New Cambridge History of India: Science, Technology, and Medicine in Colonial India* (Cambridge, UK: Cambridge University Press, 2000), 115.
14. Whitcombe, *Agrarian Conditions in Northern India*, 25.
15. From www.rainwaterharvesting.org/Rural/Traditional2.htm#Beng.
16. Baker, Dempster, and Yule, "The Prevalence of Organic Disease of the Spleen," 11.
17. H. E. Shortt and P.C.C. Garnham, "Samuel Rickard Christophers, 27 November 1873–19 February 1978," *Biographical Memoirs of Fellows of the Royal Society* 25 (November 1979): 179–207.
18. W. F. Bynum, "'Reasons for Contentment': Malaria in India, 1900–1920," *Parassitologia* 40 (1998): 19–27.
19. Sheldon Watts, "British Development Policies and Malaria in India 1897–c. 1929," *Past and Present* (November 1999): 141–81.
20. Ibid., 141–81.
21. Mridula Ramana, "Florence Nightingale and Bombay Presidency," *Social Scientist* 30, no. 9/10 (September–October 2002): 31–46.
22. Raymond E. Dumett, "The Campaign Against Malaria and the Expansion of Scientific Medical and Sanitary Services in British West Africa, 1898–1910," *African Historical Studies* 1, no. 2 (1968): 153–97.

23. J.W.W. Stephens, "Discussion on the Prophylaxis of Malaria," *British Medical Journal* (September 17, 1904).
24. Erwin H. Ackerknecht, *Malaria in the Upper Mississippi Valley, 1760–1900* (Baltimore: Johns Hopkins University Press, 1945), 5, 23, 25, 39, 44.
25. Ibid., 33–34.
26. Marie D. Gorgas and Burton J. Hendrick, *William Crawford Gorgas: His Life and Work* (New York: Doubleday, Page and Company, 1924), 41; John M. Gibson, *Physician to the World: The Life of General William C. Gorgas* (Durham, N.C.: Duke University Press, 1950), 27.
27. Gibson, *Physician to the World*, 35, 40.
28. Gorgas and Hendrick, *William Crawford Gorgas*, 6, 47; Paul Starr, *The Social Transformation of American Medicine* (New York: Basic Books, 1982), 81–85, 115–16.
29. Gibson, *Physician to the World*, 35, 43, 50; Gordon Harrison, *Mosquitoes, Malaria and Man: A History of the Hostilities Since 1880* (New York: E. P. Dutton, 1978), 158–59; David McCullough, *The Path Between the Seas: The Creation of the Panama Canal, 1870–1914* (New York: Simon & Schuster, 1977), 412.
30. Gorgas and Hendrick, *William Crawford Gorgas*, 122; Gibson, *Physician to the World*, 67; McCullough, *The Path Between the Seas*, 412, 415.
31. McCullough, *The Path Between the Seas*, 407–08, 423; Gibson, *Physician to the World*, 103, 126.
32. James Stevens Simmons, *Malaria in Panama* (Baltimore: Johns Hopkins University Press, 1939), 95; Gorgas and Hendrick, *William Crawford Gorgas*, 153; McCullough, *The Path Between the Seas*, 416.
33. Simmons, *Malaria in Panama*, 168; David A. Warrell and Herbert M. Gilles, *Essential Malariology*, 4th ed. (New York: Hodder Arnold, 2002), 331.
34. Simmons, *Malaria in Panama*, 27; McCullough, *The Path Between the Seas*, 416, 420.
35. Gibson, *Physician to the World*, 134.
36. Ibid., 105, 107, 119, 132; McCullough, *The Path Between the Seas*, 421, 452.
37. Gibson, *Physician to the World*, 123.
38. Gorgas and Hendrick, *William Crawford Gorgas*, 164.
39. McCullough, *The Path Between the Seas*, 423.
40. Ibid., 448, 451, 452, 458; Gibson, *Physician to the World*, 113–14.
41. McCullough, *The Path Between the Seas*, 467–68.
42. Dumett, "The Campaign Against Malaria," 153–97.
43. Simmons, *Malaria in Panama*, 96; Harrison, *Mosquitoes, Malaria and Man*, 166–67; McCullough, *The Path Between the Seas*, 468; Gibson, *Physician to the World*, 150.

44. Simmons, *Malaria in Panama*, 6, 36–50, 52, 56.
45. "The Real Builders of the Panama Canal," *New York Times*, October 22, 1912; B. W. Higman, "Black Labor on a White Canal: Panama, 1904–1981 (Review)," *American Historical Review* 92 (June 1987): 778.
46. "Magoon Here, Replies to Poultney Bigelow," *New York Times*, January 29, 1906, 1.
47. Michael Conniff, *Black Labor on a White Canal: Panama, 1904–1981* (Pittsburgh: University of Pittsburgh Press, 1985), 32, 38.
48. McCullough, *The Path Between the Seas*, 576–77.
49. "Panama Made Safe, Says Col. Gorgas," *New York Times*, June 13, 1907.
50. "Roosevelt Photo Gets Scant Applause," *New York Times*, November 30, 1907.
51. James L. A. Webb, *Humanity's Burden: A Global History of Malaria* (New York: Cambridge University Press, 2009), 78.
52. L. Schuyler Fonaroff, "Geographic Notes on the Barbados Malaria Epidemic," *The Professional Geographer* 18, no. 3 (May 1966): 155–63.
53. From General Gorgas's testimony in *L. D. Hand v. Alabama Power Company*, February 11, 1915, published in pamphlet form as *The Great Destroyers* by Robert L. Hughes, Anniston, Ala.
54. Conniff, *Black Labor on a White Canal*, 30.
55. McCullough, *The Path Between the Seas*, 501, 582.
56. "Gorgas's Conquest of Disease," *New York Times*, September 22, 1912, X7.
57. McCullough, *The Path Between the Seas*, 503, 587.
58. "Roosevelt Photo Gets Scant Applause."
59. Gibson, *Physician to the World*, 179, 180, 209.
60. Ibid., 165, 172, 176–77, 208.
61. Malcolm Gladwell, "The Mosquito Killer: Millions of People Owe Their Lives to Fred Soper. Why Isn't He a Hero?" *The New Yorker*, July 2, 2001.
62. Gibson, *Physician to the World*, 184, 197.
63. Henry Welles Durham, "The Clean-up of Panama," *New York Times*, July 28, 1914, 6; L. H. Woolsey, "Executive Agreements Relating to Panama," *American Journal of International Law* 37, no. 3 (July 1943): 482–89.
64. H. R. Carter, "The Effect of Variation of Level of Impounded Water on the Control of *Anopheles* Production," *Southern Medical Journal* 17, no. 8 (August 1924): 575–78.
65. Harvey H. Jackson, *Putting Loafing Streams to Work: The Building of Lay, Mitchell, Martin, and Jordan Dams, 1910–1929* (Tuscaloosa: University of Alabama Press, 1997), 40.
66. From General Gorgas's testimony in *L. D. Hand v. Alabama Power Company*.
67. Revised for clarity—original quote written phonetically. Jack Kytle, "I'm Allus Hongry," in James Seay Brown, Jr., ed., *Up Before Daylight: Life Histories from*

the Alabama Writers' Project, 1938–1939 (Tuscaloosa: University of Alabama Press, 1982), 125–26.

68. Jackson, *Putting Loafing Streams to Work*, 46.
69. Letter from T.H.D. Griffitts to Henry Rose Carter, March 12, 1913, Philip S. Hench Walter Reed Yellow Fever Collection, available at www.etext.lib.virginia.edu/etcbin/fever-browseprint?id=01022015.
70. Revised for clarity—original quote written phonetically. Kytle, "I'm Allus Hongry," 125–26.
71. W. H. Sanders, "Annual Report of the Board of Health of Alabama," Montgomery, Ala.: December 1914.
72. Harvey H. Jackson, *Putting Loafing Streams to Work*, 37, 51.
73. Interview with Harvey Jackson, August 9, 2006.
74. Revised for clarity—original quote written phonetically. Quoted in Jackson, *Putting Loafing Streams to Work*, 49, 52; Thomas Martin, *Forty Years of the Alabama Power Company, 1911–1951* (New York: Newcomen Society in North America, 1952), 13.
75. From General Gorgas's testimony in *L.D. Hand v. Alabama Power Company*.
76. *Landrift D. Hand v. Louisville Nashville Railroad Company*, Circuit Court, Shelby County, Ala., December 11, 1914.
77. Correspondence with Andrew Spielman, August 22, 2006.
78. From General Gorgas's testimony in *L.D. Hand v. Alabama Power Company*.
79. Correspondence with Andrew Spielman, October 10, 2006; from General Gorgas's testimony in *L.D. Hand v. Alabama Power Company*.
80. *Alabama Power Co. v. Carden*, Supreme Court of Alabama, 189 Ala. 384, 66 So. 596, November 7, 1914.
81. Carter, "The Effect of Variation of Level of Impounded Water," 575–78.
82. Samuel W. Welch, "Annual Report of the State Board of Health of Alabama," Montgomery, Ala., December 31, 1917.
83. "If it constructed the dam in compliance with the law it could not be guilty of negligence in causing the backing of the water," *Burnett v. Alabama Power Company*, 199 Ala. 337, 74 So. 459, December 21, 1916.
84. Theodore Steinberg, *Nature Incorporated: Industrialization and the Waters of New England* (Cambridge, UK: Cambridge University Press, 1991), 244; also Walter H. Voskuil, *The Economics of Water Power Development* (Chicago and New York: A.W. Shaw Company, 1928), 15; and Ackerknecht, *Malaria in the Upper Mississippi Valley*, 72.
85. Ann Vileisis, *Discovering the Unknown Landscape: A History of America's Wetlands* (Washington, D.C.: Island Press), 64, 67, 82.
86. Half of the animals now endangered in the United States and a third of endangered plants hail from wetland habitats. Ibid., 123, 124, 270.

87. Ackerknecht, *Malaria in the Upper Mississippi Valley*, 94.
88. See www.tva.gov/heritage/fdr/index.htm; Andrew Spielman and Michael D'Antonio, *Mosquito: A Natural History of Our Most Persistent and Deadly Foe* (New York: Hyperion, 2001), 152.
89. Margaret Humphreys, *Malaria: Poverty, Race, and Public Health in the United States* (Baltimore: Johns Hopkins University Press, 2001), 111; John Duffy, "Impact of Malaria on the South," in Todd L. Savitt and James Harvey Young, eds., *Disease and Distinctiveness in the American South* (Knoxville: University of Tennessee Press, 1988), 50.
90. Voskuil, *The Economics of Water Power Development*, 146; T.H.D. Griffitts, "Impounded Waters and Malaria," *Southern Medical Journal* 19 (1926): 367–70; Carter, "The Effect of Variation of Level of Impounded Water," 575–78.
91. See www.tva.gov/heritage/fdr/index.htm; Spielman and D'Antonio, *Mosquito*, 152.
92. Humphreys, *Malaria*, 142.
93. Mark Overton, "Agricultural Revolution in England, 1500–1850," available at www.bbc.co.uk/history.
94. Hackett, *Malaria in Europe*, 89.
95. Mark Overton, "The Diffusion of Agricultural Innovations in Early Modern England: Turnips and Clover in Norfolk and Suffolk, 1580–1740," *Transactions of the Institute of British Geographers,* New Series, 10, no. 2 (1985): 205–21.
96. E. L. Jones, "Agriculture and Economic Growth in England, 1660–1750: Agricultural Change," *The Journal of Economic History* 25, no. 1 (March 1965): 1–18.
97. Hackett, *Malaria in Europe*, 56, 63, 69.
98. Ibid., 64.
99. www.bbc.co.uk/history/british/empire_seapower/agricultural_revolution_02.shtml.
100. Letter from Dr. Livingstone to the editor, *Medical Times and Gazette*, January 26, 1863.
101. Dobson, "'Marsh Fever,'" 357–89.

9. THE SPRAY-GUN WAR

1. O. R. McCoy, "Malaria and the War," *Science* 100, no. 2607 (December 15, 1944): 535–39.
2. Mark Harrison, "Medicine and the Culture of Command: The Case of Malaria Control in the British Army During the Two World Wars," *Medical History* 40 (1996): 437–52.

3. Frank M. Snowden, *The Conquest of Malaria: Italy, 1900–1962* (New Haven, Conn.: Yale University Press, 2006), 188–89.
4. Anne O'Hare McCormick, "Undoing the German Campaign of the Mosquito," *New York Times*, September 13, 1944.
5. Snowden, *The Conquest of Malaria*, 196–97.
6. John H. Perkins, "Reshaping Technology in Wartime: The Effect of Military Goals on Entomological Research and Insect-Control Practices," *Technology and Culture* 19, no. 2 (April 1978): 169–86.
7. Christopher J. Bosso, *Pesticides and Politics: The Life Cycle of a Public Issue* (Pittsburgh: University of Pittsburgh Press, 1987), 30.
8. George W. Ware and David M. Whitacre, *The Pesticide Book*, 6th ed. (Willoughby, Ohio: MeisterPro Information Resources, 2004); "Toxicological Profile for DDT, DDE, and DDD," Agency for Toxic Substance and Disease Registry, 2002; International Programme on Chemical Safety, "Global Assessment of the State-of-the-Science of Endocrine Disruptors," World Health Organization, 2002.
9. Ware and Whitacre, *The Pesticide Book*; Edmund Russell, "The Strange Career of DDT: Experts, Federal Capacity, and Environmentalism in World War II," *Technology and Culture* 40 (1999): 770–96.
10. Bosso, *Pesticides and Politics*, 30.
11. "Public to Receive DDT Insecticide," *New York Times*, July 27, 1945.
12. Bosso, *Pesticides and Politics*, 31.
13. Perkins, "Reshaping Technology in Wartime," 169–86.
14. Waldemar Kaempeffert, "DDT, the Army's Insect Powder, Strikes a Blow Against Typhus and for Pest Control," *New York Times*, June 4, 1944.
15. Joshua Blu Buhs, *The Fire Ant Wars: Nature, Science, and Public Policy in Twentieth-century America* (Chicago: University of Chicago Press, 2004), 69.
16. E. P. Russell III, "Speaking of Annihilation: Mobilizing for War Against Human and Insect Enemies, 1914–1945," *Journal of American History* 82 (1996): 1505–29.
17. Ibid.
18. James Whorton, *Before Silent Spring: Pesticides and Public Health in Pre-DDT America* (Princeton, N.J.: Princeton University Press, 1974), 249.
19. Blu Buhs, *The Fire Ant Wars*, 68.
20. Russell, "Speaking of Annihilation," 1505–29.
21. Clay Lyle, "Achievements and Possibilities in Pest Eradication," *Journal of Economic Entomology* 40 (February 1947): 1–8.
22. Bosso, *Pesticides and Politics*, 81.
23. Quoted in Gordon Harrison, *Mosquitoes, Malaria and Man: A History of the Hostilities Since 1880* (New York: E. P. Dutton, 1978), 223.

24. John Farley, *To Cast Out Disease: A History of the International Health Division of the Rockefeller Foundation (1913–1951)* (Oxford: Oxford University Press, 2004), 143.
25. Ibid., 130.
26. John Duffy, ed., *Ventures in World Health: The Memoirs of Fred Lowe Soper* (Washington, D.C.: Pan American Health Organization, 1977), viii.
27. Malcolm Gladwell, "The Mosquito Killer: Millions of People Owe Their Lives to Fred Soper. Why Isn't He a Hero?" *The New Yorker*, July 2, 2001.
28. Farley, *To Cast Out Disease*, 144.
29. Quoted in Harrison, *Mosquitoes, Malaria and Man*, 223.
30. Peter J. Brown, "Malaria, *Miseria,* and Underpopulation in Sardinia: The 'Malaria Blocks Development' Cultural Model," *Medical Anthropology* 17 (1997): 239–54.
31. Ibid.
32. John N. Popham, "Report Progress in Malaria Fight," *New York Times*, December 7, 1948.
33. Blu Buhs, *The Fire Ant Wars*, 73; "DDT Saves the Pines," *New York Times*, September 6, 1947; "Conservation: The Menace of DDT," *New York Times*, March 1, 1959.
34. Interview with Anna Opel, March 22, 2006.
35. Sonia Shah, *Crude: The Story of Oil* (New York: Seven Stories, 2004), 18.
36. Harrison, *Mosquitoes, Malaria and Man*, 230–31; Andrew Spielman and Michael D'Antonio, *Mosquito: A Natural History of Our Most Persistent and Deadly Foe* (New York: Hyperion, 2001), 149.
37. Snowden, *The Conquest of Malaria*, 205; Harrison, *Mosquitoes, Malaria and Man*, 229; Andrew Spielman et al., "Time Limitation and the Role of Research in the World-wide Attempt to Eradicate Malaria," *Journal of Medical Entomology* 30, no. 1 (January 1993): 6–19.
38. Farley, *To Cast Out Disease*, 285.
39. Gladwell, "The Mosquito Killer."
40. Randall Packard, *The Making of a Tropical Disease: A Short History of Malaria* (Baltimore: Johns Hopkins University Press, 2007), 144.
41. "U.N. Gains Ground Against Malaria," *New York Times*, June 7, 1952.
42. M. J. Dobson et al., "Malaria Control in East Africa: The Kampala Conference and the Pare-Taveta Scheme: A Meeting of Common and High Ground," *Parassitologia* 42 (2000): 149–166.
43. Perkins, "Reshaping Technology in Wartime," 169–86.
44. Russell, "The Strange Career of DDT," 770–96.
45. Paul F. Russell, "Lessons in Malariology from World War II," *American Journal of Tropical Medicine* 26 (1946): 5–13.

46. "Public to Receive DDT Insecticide."
47. Russell, "Speaking of Annihilation," 1505–29.
48. Russell, "The Strange Career of DDT," 770–96.
49. Perkins, "Reshaping Technology in Wartime," 169–86.
50. Whorton, *Before Silent Spring*, 251.
51. Bosso, *Pesticides and Politics*, 63.
52. Perkins, "Reshaping Technology in Wartime," 169–86.
53. "Flies Resist DDT," *New York Times*, October 31, 1948.
54. WHO Expert Committee on Malaria, 1947, "Report on Dr. Pampana's Mission to Greece and Italy," WHO Docs., IC/Mal/8/21, August 1947, quoted in Harrison, *Mosquitoes, Malaria and Man*, 233.
55. "Flies Resist DDT."
56. Popham, "Report Progress in Malaria Fight."
57. Paul F. Russell, *Man's Mastery of Malaria* (London: Oxford University Press, 1955), 148.
58. Greer Williams, *The Plague Killers: Untold Stories of Three Great Campaigns Against Disease* (New York: Charles Scribner's Sons, 1969), 175.
59. J. A. Nájera, "Malaria and the Work of the WHO," *Bulletin of the World Health Organization* 67, no. 3 (1989): 229–43.
60. Ibid.
61. Socrates Litsios, "Malaria and International Health Organizations," prepared for "Philanthropic Foundations and the Globalization of Scientific Medicine," Quinnipiac University, November 6–18, 2003, quoting WHO, Eighth World Health Assembly, Mexico, May 10–27, 1955. Official Records of the WHO, No. 63: 205.
62. James L. A. Webb, *Humanity's Burden: A Global History of Malaria* (New York: Cambridge University Press, 2009), 167; "Soviet Aid Offer on Malaria Cited," *New York Times*, January 12, 1958.
63. R. M. Packard, "'No Other Logical Choice': Global Malaria Eradication and the Politics of International Health in the Post-war Era," *Parassitologia* 40 (1998), 217–29.
64. "Malaria Eradication," Report and Recommendations of the International Development Advisory Board, Washington, D.C., April 13, 1956.
65. Harry Cleaver, "Malaria and the Political Economy of Public Health," *International Journal of Health Services* 7, no. 4 (1977): 557–79.
66. "Malaria Eradication."
67. Ibid.
68. Ibid.
69. G. Sambasivan, "Roundtable Discussion: WHO's Passive Role," *World Health Forum* 1, nos. 1, 2 (1980): 8–33.

70. "Mosquitoes Developing an Armor Against DDT After 9-year War," *New York Times*, March 14, 1952.
71. Robert K. Plumb, "Dichloro-diphenyl-trichloroethane," *New York Times*, January 16, 1955.
72. Spielman and D'Antonio, *Mosquito*, 150; Laurie Garrett, *The Coming Plague: Newly Emerging Diseases in a World Out of Balance* (New York: Penguin Books, 1994), 50; C. P. Gilmore, "Malaria Wins Round 2," *New York Times*, September 25, 1966; Jean Mouchet, "Agriculture and Vector Resistance," *Insect Science and Its Application* 9, no. 3 (1988): 297–302; Harrison, *Mosquitoes, Malaria and Man*, 295; U. D'Alessandro and H. Buttiëns, "History and Importance of Antimalarial Drug Resistance," *Tropical Medicine and International Health* 6, no. 11 (November 2001): 845–48.
73. William G. Brogdon and Janet C. McAllister, "Insecticide Resistance and Vector Control," *Emerging Infectious Diseases* 4, no. 4 (October–December 1998): 605–13.
74. Brown, "Malaria, *Miseria,* and Underpopulation in Sardinia," 239–54.
75. Willard H. Wright, *Forty Years of Tropical Medicine Research: A History of the Gorgas Memorial Institute of Tropical and Preventive Medicine, Inc., and the Gorgas Memorial Laboratory* (Baltimore: Reese Press, 1970), 102.
76. In fact, the problem of resistant mosquitoes wasn't quite as straightforward as Russell and IDAB suggested. Under a DDT onslaught, any mosquito that could avoid fatal DDT poisoning had a strong advantage, true. The ones that could imperviously absorb the compound and fly away were truly dangerous, since they could both resist the DDT and effectively transmit malaria. But there were others. Many *Anopheles* mosquitoes found DDT-treated surfaces irritating and simply avoided them altogether, opting to bite or rest, or both, out of doors, free from the enervating chemical. In one study, as many as three out of five mosquitoes actively avoided DDT-treated surfaces. Theoretically these DDT-repelled mosquitoes had just as much chance of mothering battalions of progeny as the DDT-resistant ones, but unlike the resistant mosquitoes, their survival tactic allowed them much less opportunity to bite and infect human beings with malaria. In other words, even if DDT failed to kill sufficient numbers of mosquitoes, it could still repel sufficient numbers to effect the same result: a cessation of malaria transmission. "Malaria eradication," Report and Recommendations of the International Development Advisory Board, Washington, D.C., April 13, 1956.
77. Williams, *The Plague Killers*, 175–76; Spielman et al., "Time Limitation and the Role of Research," 6–19.
78. Dwight D. Eisenhower, "Annual Message to the Congress on the State of the

Union," January 9, 1958, via the American Presidency Project, www.americanpresidency.org.

79. Howard A. Rusk, "Aiding Fight on Malaria," *New York Times*, December 15, 1957.
80. "U.S. Will Help India Eradicate Malaria," *New York Times*, December 6, 1957.
81. M. A. Farid, "The Malaria Programme: From Euphoria to Anarchy," *World Health Forum* 1, no. 1 (1980): 8–33; "U.S. Will Help India Eradicate Malaria."
82. Robert S. Desowitz, *The Malaria Capers: More Tales of Parasites and People, Research and Reality* (New York: W. W. Norton, 1991), 214; M.G. Candau, "World Acts to Combat Malaria," *New York Times*, March 19, 1960.
83. Packard, *The Making of a Tropical Disease*, 157.
84. "WHO Reports 11 Countries Have Eradicated Malaria," *New York Times*, January 24, 1960; Candau, "World Acts to Combat Malaria."
85. Desowitz, *The Malaria Capers*, 14; Amy Yomiko Vittor et al., "The Effect of Deforestation on the Human-biting Rate of *Anopheles darlingi*, the Primary Vector of *Falciparum* Malaria in the Peruvian Amazon," *American Journal of Tropical Medicine and Hygiene* 74, no. 1 (2006): 3–11.
86. Harrison, *Mosquitoes, Malaria and Man*, 242–43; Gilmore, "Malaria Wins Round 2."
87. Julian de Zulueta and François Lachance, "A Malaria-Control Experiment in the Interior of Borneo," *Bulletin of the World Health Organization* 15 (1956): 673–93.
88. Brown, "Malaria, *Miseria,* and Underpopulation in Sardinia," 239–54.
89. Gilmore, "Malaria Wins Round 2."
90. De Zulueta and Lachance, "A Malaria-Control Experiment in the Interior of Borneo," 673–93.
91. Spielman and D'Antonio, *Mosquito*, 172; "Dr. Paul F. Russell, 89; Specialist on Malaria," *New York Times*, November 9, 1983.
92. Garrett, *The Coming Plague*, 49.
93. T. H. Weller, "World Health in a Changing World," *Journal of Tropical Medicine and Hygiene* 77, suppl. (April 1974): 54–61.
94. Geoffrey M. Jeffery, "Malaria Control in the Twentieth Century," *American Journal of Tropical Medicine and Hygiene* 24, no. 3 (1976): 361–71.
95. F. Y. Cheng, "Deterioration of Thatch Roofs by Moth Larvae after House Spraying in the Course of a Malaria Eradication Programme in North Borneo," *Bulletin of the World Health Organization* 28 (1963): 136–37.
96. Gordon R. Conway, "Ecological Aspects of Pest Control in Malaysia," from J. T. Farvar and J. P. Milton, eds., *The Careless Technology: Ecology and International Development* (New York: Natural History Press, 1972).

97. Arthur Brown, "Personal Experiences in the Malaria Eradication Campaign, 1955–1962," *Journal of the Royal Society of Medicine* 95, no. 3 (March 2002): 154–56.
98. D. K. Visnawathan, *The Conquest of Malaria in India: An Indo-American Co-operative Effort* (Madras, India: Company Law Institute Press, 195), cited in Harrison, *Mosquitoes, Malaria and Man*, 241.
99. Tom Harrisson, "Operation Cat-drop," *Animals* 5 (1965): 512–13.
100. R.A.F. *Operations Record Book*, Changi, Singapore, March 1960.
101. R. M. Packard, "Malaria Dreams: Postwar Visions of Health and Development in the Third World," *Medical Anthropology* 17 (1997): 179–96.
102. Visnawathan, *The Conquest of Malaria in India*, 195, cited in Harrison, *Mosquitoes, Malaria and Man*, 241.
103. T. C. Boyle wrote: "You should have seen them . . . the little parachutes and harnesses we'd tricked up, 14,000 of them, cats in every color of the rainbow, cats with one ear, no ears, half a tail, three-legged cats . . . all of them twirling down out of the sky like great big oversized snowflakes." Boyle's fictional fourteen thousand figure found its way into retellings of the Borneo story and was replicated widely. "As wonderful and touching and Disney-like this story is, it actually has nothing to do with the World Health Organization . . . The story has been reproduced and published in books, journals, and on the Internet, without any supporting evidence or references," huffed the WHO librarian Thomas Allen when asked about it. "Once this cat got out of the bag, there was no getting it back in." T. C. Boyle, "Top of the Food Chain," in *Without a Hero: Stories* (New York: Viking, 1994); Carole Modis, "Operation Cat Drop," *Quarterly News of the Association of Former WHO Staff* (April–June 2005); correspondence with Thomas Allen, January 15, 2008.
104. Harrison, *Mosquitoes, Malaria and Man*, 244.
105. Quoted in Gladwell, "The Mosquito Killer."
106. Harrison, *Mosquitoes, Malaria and Man*, 254.
107. "DDT: Its Days as a Killer Are Numbered," *New York Times*, November 16, 1969.
108. Nájera, "Malaria and the Work of the WHO," *Bulletin of the World Health Organization* 67, no. 3 (1989): 229–43.
109. Harrison, *Mosquitoes, Malaria, and Man*, 295.
110. Spielman and D'Antonio, *Mosquito*, 161.
111. Harrison, *Mosquitoes, Malaria and Man*, 248.
112. G. Davidson and A. R. Zahar, "The Practical Implications of Resistance of Malaria Vectors to Insecticides," *Bulletin of the World Health Organization* 49 (1973): 475–83.
113. Mario Pinotti, "Chemoprophylaxis of Malaria by the Association of an

Antimalarical [*sic*] Drug to the Sodium Chloride Used Daily in the Preparation of Meals," *International Congresses of Tropical Medicine and Malaria* 2 (1953): 248; Rostan de Rohan Loureiro Soares, "Sal Chloroquinado, Novo Metodo Deprofilaxia da Malaria," *Revista Brasiliera de Medicina* (July 1955): 448.

114. D. Payne, "Did Medicated Salt Hasten the Spread of Chloroquine Resistance in *Plasmodium falciparum*?" *Parasitology Today* 4, no. 4 (1988).
115. D'Alessandro and Buttiëns, "History and Importance of Antimalarial Drug Resistance."
116. Howard A. Rusk, "Health Projects Abroad," *New York Times*, September 22, 1963.
117. Packard, "Malaria Dreams," 179–96.
118. Brown, "Malaria, *Miseria,* and Underpopulation in Sardinia," 239–54.
119. Farley, *To Cast Out Disease*, 291.
120. Brown, "Malaria, *Miseria,* and Underpopulation in Sardinia," 239–54.
121. M. A. Farid, "The Malaria Programme: From Euphoria to Anarchy," *World Health Forum* 1, no. 1 (1980): 8–33.
122. Farley, *To Cast Out Disease*, 277.
123. Packard, *The Making of a Tropical Disease*, 147.
124. Farley, *To Cast Out Disease*, 273, citing T. Poleman, "World Food: A Perspective," *Science* 188 (1975): 510–28, and Kingsley Davis, "The Population Specter: Rapidly Declining Death Rate in Densely Populated Countries," *American Economic Review* 46 (1956): 305–18.
125. Williams, *The Plague Killers*, 181.
126. Russell, *Man's Mastery of Malaria*, 246.
127. Ware and Whitacre, *The Pesticide Book*; "Toxicological Profile for DDT, DDE, and DDD," Agency for Toxic Substance and Disease Registry, 2002; International Programme on Chemical Safety, "Global Assessment of the State-of-the-Science of Endocrine Disruptors."
128. "Conservation: The Menace of DDT," *New York Times*, March 1, 1959.
129. "Farmers Warned on DDT," *New York Times*, May 25, 1947.
130. Plumb, "Dichloro-diphenyl-trichloroethane."
131. Ralph H. Lutts, "Chemical Fallout: *Silent Spring*, Radioactive Fallout, and the Environmental Movement," in Craig Waddell, ed., *And No Birds Sing: Rhetorical Analyses of Rachel Carson's* Silent Spring (Carbondale: Southern Illinois University Press, 2000), 24.
132. Terence Kehoe and Charles Jacobson, "Environmental Decision Making and DDT Production at Montrose Chemical Corporation of California," *Enterprise and Society* 4, no. 4 (2003); "U.S. Seeks to Keep Milk Free of DDT," *New York Times*, April 23, 1949.

133. Bosso, *Pesticides and Politics*, 122–23.
134. Webb, *Humanity's Burden*, 172.
135. Packard, *The Making of a Tropical Disease*, 169.
136. Weller, "World Health in a Changing World," 54–61.
137. Spielman and D'Antonio, *Mosquito*, 173.
138. Harrison, *Mosquitoes, Malaria, and Man*, 248, 253.
139. L. J. Bruce-Chwatt, "Resurgence of Malaria and Its Control," *Journal of Tropical Medicine and Hygiene* 77, suppl. (April 1974): 62–66.
140. Ibid.
141. Socrates Litsios, *The Tomorrow of Malaria* (Wellington, New Zealand: Pacific Press, 1996), 101.
142. Harrison, *Mosquitoes, Malaria, and Man*, 257.
143. Jonathan A. Leonard, "Malaria Strikes Back: A 'Dying' Disease Kills Again," *Chicago Tribune*, August 15, 1979.
144. Nájera, "Malaria and the Work of the WHO," 229–43.
145. Alan Riding, "Malaria Spreading in Central America as Resistance to Sprays Grows," *New York Times*, August 23, 1977.
146. Spielman and D'Antonio, *Mosquito*, 172.
147. Packard, *The Making of a Tropical Disease*, 174.
148. Historical currency conversion from 1960 dollars to 2009 dollars calculated with www.futureboy.homeip.net/fsp/dollar.fsp?quantity=7¤cy=dollars&fromYear=1960.
149. José Nájera, "Malaria: New Patterns and Perspectives," World Bank Technical Paper Number 183 (Washington, D.C.: World Bank, 1992).
150. Harrison, *Mosquitoes, Malaria, and Man*, 296.
151. Packard, *The Making of a Tropical Disease*, 159.
152. Nájera, "Malaria: New Patterns and Perspectives."
153. Elisabeth Rosenthal, "Outwitted by Malaria, Desperate Doctors Seek New Remedies," *New York Times*, February 12, 1991.
154. Walther H. Wernsdorfer, "Epidemiology of Drug Resistance in Malaria," *Acta Tropica* 56 (1994): 143–56.
155. Harrison, *Mosquitoes, Malaria, and Man*, 295.
156. www.who.int/mediacentre/factsheets/smallpox/en/.
157. Andrew Spielman et al., "Time Limitation and the Role of Research in the Worldwide Attempt to Eradicate Malaria," *Journal of Medical Entomology* 30, no. 1 (January 1993): 6–19.
158. Sambasivan, "Roundtable Discussion: WHO's Passive Role," 8–33.
159. Georgann Chapin and Robert Wasserstrom, "Agricultural Production and Malaria Resurgence in Central America and India," *Nature* 293 (September 17, 1981): 181–85.

160. Andrew Spielman, lecture at Harvard University, March 2, 2006.
161. A. P. Ray, "Roundtable Discussion: Warning Should Be Heeded," *World Health Forum* 1, nos. 1, 2 (1980): 8–33.
162. Paul F. Russell, "Roundtable Discussion: Goal of Eradication Must Be Maintained," *World Health Forum* 1, nos. 1, 2 (1980): 8–33.
163. Robert H. Black, "Roundtable Discussion: Farid Is Right," *World Health Forum* 1, nos. 1, 2 (1980): 8–33.
164. Email correspondence with Donald Roberts, April 4, 2006.

10. THE SECRET IN THE MOSQUITO

1. Jason L. Riley, "Malaria's Toll," *Wall Street Journal*, August 21, 2006.
2. M. J. Dobson et al., "Malaria Control in East Africa: The Kampala Conference and the Pare-Taveta Scheme: A Meeting of Common and High Ground," *Parassitologia* 42 (2000): 149–166.
3. Interview with Amir Attaran, February 9, 2006; Andrew Spielman and Michael D'Antonio, *Mosquito: A Natural History of Our Most Persistent and Deadly Foe* (New York: Hyperion, 2001), 166.
4. Quoted in Jeffrey Sachs, *The End of Poverty: Economic Possibilities for Our Time* (New York: Penguin Books, 2005), 190.
5. Meredith Fort, Mary Anne Mercer, Oscar Gish, eds., *Sickness and Wealth: The Corporate Assault on Global Health* (Cambridge, Mass.: South End Press, 2004), 205, and Jim Yong Kim, Joyce V. Millen, Alec Irwin, and John Gershman, eds., *Dying for Growth: Global Inequality and the Health of the Poor* (Monroe, Maine: Common Courage Press, 2000), 143.
6. Sarah Sexton, "Trading Health Care Away? GATS, Public Services and Privatization," Corner House briefing 23, July 2001.
7. Badria Babiker El Sayed, et al., "A Study of the Urban Malaria Transmission Problem in Khartoum," *Acta Tropica* 75 (2000): 163–71.
8. Socrates Litsios, *The Tomorrow of Malaria* (Wellington, N.Z.: Pacific Press, 1996), 127.
9. www.un.org/esa/population/publications/adultmort/UNAIDS_WHOPaper2.pdf.
10. Laith J. Abu-Raddad et al., "Dual Infection with HIV and Malaria Fuels the Spread of Both Diseases in Sub-Saharan Africa," *Science* 314, no. 5805 (December 8, 2006): 1603–606.
11. Randall Packard, *The Making of a Tropical Disease: A Short History of Malaria* (Baltimore: Johns Hopkins University Press, 2007), 217.
12. H. Kristian Heggenhougen et al., *The Behavioural and Social Aspects of Malaria and Its Control* (World Health Organization, 2003), 87.

13. Press Release, "Dr. Gro Harlem Brundtland Elected Director-General of the World Health Organization," May 15, 1998.
14. Mark Grabowsky, "The Billion-Dollar Malaria Moment," *Nature* 451 (February 28, 2008): 1051–52.
15. Jenny Anderson, "Fighting a Disease of Logistics, He Means Business," *New York Times*, November 12, 2007.
16. Juhie Bhatia, "Twitter Face-off to Fight Malaria," Global Voices Online, April 20, 2009, www.globalvoicesonline.org/2009/04/20/global-health-twitter-face-off-to-fight-malaria/.
17. See www.americanidol.com/idolgivesback/.
18. Mark Honigsbaum, "Net Effects," *The Guardian*, April 24, 2007.
19. Donald G. McNeil, "A $10 Mosquito Net Is Making Charity Cool," *New York Times*, June 2, 2008.
20. Suzanne Malveaux, "'Idol' Star Boosts First Lady's Anti-malaria Event in Africa," CNN.com, July 3, 2007.
21. Packard, *The Making of a Tropical Disease*, 223.
22. Rosanne Skirble, "Economic Downturn Threatens Global Fund for AIDS, TB, Malaria," VOANews.com, February 4, 2009.
23. McNeil, "A $10 Mosquito Net."
24. Beth Gorham, "Belinda Stronach Joins Heavyweights at Washington Gathering on Malaria," Canadian Press, December 13, 2006.
25. Malveaux, "'Idol' Star Boosts First Lady's Anti-malaria Event in Africa."
26. McNeil, "A $10 Mosquito Net."
27. 2008 Millennium Development Goals Malaria Summit, September 25, 2008, New York. Video footage at www.dfid.gov.uk/news/files/malaria-summit.asp.
28. Quoted in L. J. Bruce-Chwatt, "Paleogenesis and Paleo-epidemiology of Primate Malaria," *Bulletin of the World Health Organization* 32 (1965): 376.
29. Ronald Ross, *Memoirs* (London: John Murray, 1928), 448.
30. Lewis W. Hackett, *Malaria in Europe: An Ecological Study* (London: Oxford University Press, 1937), 294.
31. M. Eveillard et al., "Measurement and Interpretation of Hand Hygiene Compliance Rates: Importance of Monitoring Entire Care Episodes," *Journal of Hospital Infection* 72, no. 3 (May 28, 2009): 211–17.
32. Heggenhougen et al., *The Behavioural and Social Aspects of Malaria and Its Control*, 87, 102.
33. "Information Gap Challenges Zanzibar's Antimalaria Campaign," AllAfrica.com, May 12, 2006.
34. Heggenhougen et al., *The Behavioural and Social Aspects of Malaria and Its Control*, 94–95.

35. Philip Adongo, "How Local Community Knowledge About Malaria Affects Insecticide-Treated Net Use in Northern Ghana," November 15, 2005; and Soori Nnko, "Public Health Campaigns' Dilemma: Field Experience About Malaria Control in a Rural Setting, North-western Tanzania," November 15, 2005.
36. Anna Ingwafa, "Kamwi Warns on Abuse of Mosquito Nets," AllAfrica.com, March 13, 2008.
37. Heggenhougen et al., *The Behavioural and Social Aspects of Malaria and Its Control*, 103.
38. Koremromp El et al., "Monitoring Mosquito Net Coverage for Malaria Control in Africa: Possession vs. Use by Children Under 5 Years," *Tropical Medicine and International Health* 8, no. 8 (August 2003): 693–703.
39. William Takken, "Do Insecticide-treated Bednets Have an Effect on Malaria Vectors?" *Tropical Medicine and International Health* 7, no. 12 (December 2002): 1022–30.
40. "Pyrethrins: Bright Signs After Washout Last Year," *Chemical Week*, January 17, 1979, 42.
41. F. Chandre et al., "Status of Pyrethroid Resistance in *Anopheles gambiae* Sensu Lato," *Bulletin of the World Health Organization* 77, no. 3 (1999); interview with John Thomas, December 12, 2005; and interview with Willem Takken, November 14, 2005; also David Firn, "How Syngenta Went Against the Grain and Grew," *Financial Times*, February 19, 2004, 10.
42. Abdoulaye Diabate et al., "The Role of Agricultural Use of Insecticides in Resistance to Pyrethroids in *Anopheles gambiae* s.l. in Burkina Faso," *American Journal of Tropical Medicine and Hygiene* 67, no. 6 (2002): 617–22.
43. Morteza Zaim and Pierre Guillet, "Alternative Insecticides: An Urgent Need," *Trends in Parasitology* 18, no. 4 (April 2002): 161; Chandre et al., "Status of Pyrethroid Resistance in *Anopheles gambiae* Sensu Lato."
44. Josiane Etang, Fourth MIM Pan-African Malaria Conference, November 2005, Yaoundé, Cameroon; also author correspondence with Josiane Etang, November 23, 2005.
45. www.politico.com/news/stories/0309/20160.html.
46. Oliver Sabot, "Getting to Zero: A New Global Malaria Control and Elimination Strategy," First Yale International Symposium, "The Global Crisis of Malaria: Lessons of the Past and Future Prospects," November 7–9, 2008, New Haven, Conn.
47. First Yale International Symposium, "The Global Crisis of Malaria."
48. Andrew Spielman lecture, Harvard University, March 2, 2006.
49. Awash Teklehaimanot et al., "Coming to Grips with Malaria in the New Mil-

lennium," UN Millennium Project Task Force on HIV/AIDS, Malaria, TB and Access to Essential Medicines Working Group on Malaria, 2005, 6.

50. Laura Blue, "Global Malaria Estimates Are Reduced," Time.com, September 18, 2008.
51. See www.unicef.org/statistics/index_24302.html.
52. Andrew Spielman lecture, Harvard University, March 2, 2006.
53. M. Ettling et al., "Economic Impact of Malaria in Malawian Households," *Tropical Medicine and Parasitology* 45 (1994): 74–79.
54. Jeffrey Sachs, "Power of One: The $10 Solution," *Time*, January 4, 2007.
55. Sonia Shah, *Crude: The Story of Oil* (New York: Seven Stories Press, 2004), 53.
56. Sebastian Junger, "Enter China, the Giant," *Vanity Fair*, July 2007, 126–38.
57. Christine Gorman, "Marathon Fights Malaria," *Time*, August 20, 2006.
58. Andrew Spielman et al., "Industrial Anti-Malaria Policies," Center for International Development, Harvard University, Cambridge, Mass., 2002.
59. Sharon LaFraniere, "Business Joins African Effort to Cut Malaria," *New York Times*, June 29, 2006.
60. Ibid.
61. See www.marathon.com/Social_Responsibility/Making_a_Difference/Malaria_Control_Project/.
62. Ibid.
63. "We're Making Sure Children Have a Future. And Malaria Doesn't," Advertisement, Marathon Oil, *New York Times*, April 25, 2007.
64. Eric Rezsnyak, "'American Idol' 2009: The Judges Choose to Go Insane," *Rochester City Newspaper*, May 12, 2009.
65. Shah, *Crude*, 157.
66. Sachs, "Power of One."
67. Michael Fletcher, "Bush Has Tripled Aid to Africa," *Washington Post*, December 31, 2006.
68. Roger Bate, "The Rise, Fall, Rise and Imminent Fall of DDT," *Health Policy Outlook*, November 2007.
69. Kirsten Weir, "Rachel Carson's Birthday Bashing," Salon.com, June 29, 2007.
70. Interview with Tom McCutchan, October 9, 2008.
71. Jason L. Riley, "Malaria's Toll," *Wall Street Journal*, August 21, 2006.
72. Kirsten Weir, "Rachel Carson's Birthday Bashing."
73. "Bush Announces Initiative to Fight Malaria in Africa," press release, June 30, 2005.
74. Interview with Chris Hentschel, October 16, 2008.
75. Roger Bate, "Funding Isn't Everything," *The American*, February 28, 2008.
76. First Yale International Symposium, "The Global Crisis of Malaria."

77. World Health Organization, "Report of the Technical Expert Group (TEG) Meeting on Intermittent Preventive Therapy in Infancy (IPTI)," October 8–10, 2007.
78. Comments by Brian Greenwood, First Yale International Symposium, "The Global Crisis of Malaria."
79. "Assessment of the Role of Intermittent Preventive Treatment for Malaria in Infants: Letter Report," Committee on the Perspectives on the Role of Intermittent Preventive Treatment for Malaria in Infants, 2008.
80. Donald G. McNeil, "An Iron Fist Joins the Malaria Wars," *New York Times*, June 27, 2006.
81. Donald G. McNeil, "Gates Foundation's Influence Criticized," *New York Times*, February 16, 2008.
82. Comments by Brian Greenwood, First Yale International Symposium, "The Global Crisis of Malaria"; also interview with Robert Ridley, December 11, 2008.
83. First Yale International Symposium, "The Global Crisis of Malaria."
84. Interview with Chris Hentschel, October 16, 2008.
85. "Eradicate Malaria? Doubters Fuel Debate," *New York Times*, March 4, 2008.
86. Tom Paulson, "WHO Chief Joins Gateses' Call to Eradicate Malaria," *Seattle Post-Intelligencer*, October 17, 2007.
87. Richard G. A. Feachem and Allison A. Phillips, "Malaria: 2 Years in the Fast Lane," *The Lancet* 373, no. 9673 (April 2009): 1409–11.
88. 2008 Millennium Development Goals Malaria Summit, September 25, 2008, New York. Video footage at www.dfid.gov.uk/news/files/malaria-summit.asp.
89. A. Bosman and K. N. Mendis, "A Major Transition in Malaria Treatment: The Adoption and Deployment of Artemisinin-Based Combination Therapies," *American Journal of Tropical Medicine and Hygiene* 77, suppl. (December 2007): 193–97.
90. Abdisalan M. Noor et al., "Insecticide-treated Net Coverage in Africa: Mapping Progress in 2000–07," *The Lancet* 373, no. 9657 (November 28, 2008): 58–67.
91. Interview with Thomas Ritchie, First Yale International Symposium, "The Global Crisis of Malaria."
92. Dagi Kimani, "Coartem Under Fire as $8 Million Stocks Arrive," *The East African*, May 29, 2006.
93. Interview with Edugie Abebe, Fourth MIM Pan-African Malaria Conference, November 15, 2005, Yaoundé, Cameroon.
94. Socrates Litsios, "Malaria and International Health Organizations," prepared for "Philanthropic Foundations and the Globalization of Scientific Medicine," Quinnipiac University, November 6–18, 2003.

95. Miriam K. Laufer et al., "Return of Chloroquine Antimalarial Efficacy in Malawi," *New England Journal of Medicine* 355, no. 19 (November 9, 2006): 1959–65.
96. Interview with Tom McCutchan, October 9, 2008.
97. "Monkey Malaria More Widespread in Humans: Study," Reuters, January 18, 2008.
98. Interview with Vicente Bayard, Gorgas Institute, April 21, 2006.
99. See www.cdc.gov/ncidod/dvbid/westnile/surv&control05Maps.htm and www.cdc.gov/ncidod/dvbid/westnile/surv&controlCaseCount08_detailed.htm.
100. Sonja Mali et al., "Malaria Surveillance—United States, 2007," *Morbidity and Mortality Weekly Report* 58, no. 55-2 (April 17, 2009).
101. See www.cdc.gov/ncidod/EID/vol2no1/zuckerei.htm.
102. B. Doudier et al., "Possible Autochthonous Malaria from Marseille to Minneapolis," *Emerging Infectious Diseases* 13, no. 8 (August 2007): 1236–38; and A. Krüger et al., "Two Cases of Autochthonous *Plasmodium falciparum* Malaria in Germany with Evidence for Local Transmission by Indigenous *Anopheles plumbeus*," *Tropical Medicine and International Health* 6, no. 12 (December 2001): 983–85.

ACKNOWLEDGMENTS

Most every malariologist I approached in connection with this book was patient, generous, and forthright. Among them, Terrie Taylor in Malawi, José Calzada in Panama City, and the late Andy Spielman of Harvard University were especially so. I thank them heartily. For their enthusiastic research assistance, thanks to Peter Ross, Lukas Rieppel, Mónica García, Emily Tucker, and Annie Jack; for early financial assistance, the Nation Institute and the Puffin Foundation; for their tips and their support, Carolyn and David Bulmer, Loie Hayes and Julie Ogletree, Darwin Marcus Johnson, Brian King, Hasmukh and Hansa Shah, and Susy Wasiak.

Much gratitude goes to Sarah Crichton, whose editorial acumen greatly improved this book, and to the malaria experts Wallace Peters, Prabhjot Singh, Malcolm Molyneux, and Arba Ager, who generously helped improve its accuracy. And to my agents, Charlotte Sheedy and Anthony Arnove, who supported this book from the beginning. Without them, it would not have been written.

Finally, I thank Mark, Zakir, and Kush Bulmer for sustaining me through years of researching and writing about malaria, suffering many mosquito bites along the way.

ACKNOWLEDGMENTS

INDEX

Ackerknecht, Erwin, 56
Adongo, Philip, 226
Adrian VI, Pope, 68
Aedes mosquitoes, 51; *A. aegypti*, 180
Afghanistan, 8, 216
Africa Fighting Malaria, 233
African slaves, 41–48, 52–53, 57, 62, 90, 91, 138, 177, 183
Agency for International Development, U.S. (USAID), 115, 167, 215
agriculture, 79–81, 189–92, 213, 214; in Africa, 36; drainage of wetlands for, 190; in India, 175; in Italy, 70; pesticide use in, 211, 217
Agriculture, U.S. Department of (USDA), 194, 195, 197, 198, 200, 214
AIDS, 222, 236
Alabama, 57, 58, 178, 185–89; during Civil War, 74; Supreme Court of, 188; University of, 185
Alabama Power Company, 186, 279*n83*
Alaric the Visigoth, 67
Alexander VI, Pope, 68
Algeria, 144
Algonquins, 39
Allen, Thomas, 286*n103*
Al Qaeda, 232
Amazon, 80–81
American colonies: English, 38–42, 46–48, 249*n19*; Spanish, 39, 42, 45–46, 48–53
American Enterprise Institute, 233
American Idol (television program), 224, 232
American Medical Association (AMA), 185
American Red Cross, 260*n45*
American Revolution, 71, 76, 90–91, 177
Andersen, Hans Christian, 70
Angola, 212, 233
Anopheles mosquitoes, 15, 16, 44–45, 64, 101, 162–64, 225, 241, 253*n8*; DDT and, 199, 201, 209, 211, 217, 227; in England, 171–73; Gorgas on, 188; habitats of, 44, 60–61, 63, 81,

Anopheles mosquitoes (*cont.*)
189; identification as malaria vector of, 155–60
species: *A. albimanus*, 180; *A. albitarsis*, 180; *A. aquasalis*, 180; *A. arabienses*, 25, 84; *A. atroparvus*, 61–63, 65; *A. crucians*, 39; *A. darlingii*, 180; *A. dirus*, 110; *A. farauti*, 84; *A. gambiae*, 25–26, 36, 59, 123, 125, 131–32, 137, 168, 171, 197, 199, 227, 235; *A. labranchiae*, 62–63, 65, 68, 163, 194, 198; *A. maculipennis*, 77, 158, 163, 192; *A. messeae*, 163; *A. punctimacula*, 180; *A. punctipennis*, 39, 70–72, 164, 185, 186; *A. quadrimaculatus*, 39, 71, 72, 76, 164, 186, 188; *A. rossi*, 160; *A. stephensi*, 110; *A. subpictus*, 160; *A. superpictus*, 77
Arenco Pharmaceuticals, 116–17
Arizona, University of, 65–66
Armed Forces Radio, 100
Army, U.S., 99–101, 145, 185, 188, 195
artemether-lumefantrine, 114
Artemisia annua, 112
artemisinin, 112–20, 168, 223; combination therapy (ACT), 115–16, 118, 120, 139, 237, 238, 267*n195*
Asante Empire, 89
Asian Malaria Conference (1954), 202
Atebrin, *see* quinacrine
Atlantic languages, 26
Attaran, Amir, 116
Australia, 38, 83, 228; in World War II, 99, 104–107
avian flu, 9
Azerbaijan, 8
Bacillus malariae, 144–46, 151
Bank of England, 48
Bantu peoples, 25, 35–36, 64
Barbados, 46, 183, 184
Bass, Willie, 186
Bataan, 99–100, 194
Bate, Roger, 233
Beatty, Alfred Chester, 80
bed nets, 115, 122, 124–25, 139, 231, 239; insecticide-treated, 223–28
Begum, Shahida, 171
Bethesda Naval Hospital, 108
Bignami, Amico, 154–57
Bill and Melinda Gates Foundation, 167, 223, 234, 235
Billiton Mining, 231–32
Binka, Fred, 114, 116
bin Ladin, Osama, 232
Bioko Island, 56, 231, 232
Blanton, Wyndham, 249*n19*
Bloland, Peter, 109
Boardman, Elijah, 71–73, 76, 90
Boardman, William, 72–73
Bolivia, 83, 91
Bono, 223, 236
Borland, Francis, 55–56
Borneo, 207, 209–10, 286*n103*
Boyle, T. C., 210, 286*n103*
Bradford, William, 43
Brazil, 80–81, 83, 197, 212, 217; colonial, 42
Bridgeland, John, 225
Britain, 96, 147–49, 151, 155, 172–73, 184, 189, 191–92; African colonies of, 38, 89, 174–75; American colonies of, 38–43, 46–48; during Little Ice Age, 83; Parliament, 170–73, 192; slave trade opposed by, 90–91, 138; in World War I, 77, 79,

175; in World War II, 194; *see also* India, British Raj in
British Broadcasting Company (BBC), 136
British Medical Association, 156, 157
British Medical Journal, 149
Brown, Gordon, 223
Brown, Peter, 213
Bruce-Chwatt, Leonard, 133
Brundtland, Gro, 223
Brunei, 203
Buddhists, 209, 210
Buel, Dr., 73
Bulgaria, 77, 78
Burma, 116
Burnett v. Alabama Power Company (1916), 279*n83*
Bush, George W., 223, 227–28, 232–34
Buxton, Thomas Fowell, 38, 91

Cable News Network (CNN), 225
Caesar, Julius, 65
Calcutta Medical School, 95
Calzada, José, 5–9
Cambodia, 108, 118, 120, 208, 209, 212, 267*n195*
Cameroon, 14, 34, 56, 86, 128–29, 136, 220, 227
Canada, 50
cancer, 129
Candau, Marcolino, 199
Caracalla, Emperor, 64
Carolinas, colonial, 46–48, 56; Scottish settlers in, 46, 50
Carroll, Dennis, 115
Carson, Rachel, 215, 233
Carter, Richard, 13
Caventou, Joseph, 91, 95, 102–103
Celsus, 64
Centers for Disease Control (CDC), 109, 115, 191, 240
Central African Republic, 227
Central Intelligence Agency (CIA), 221
cerebral malaria, 127, 129, 231
Chad, 231
Chagres fever, 150
Chamberlain, Joseph, 174
Chan, Margaret, 236
Chaouch, Adel, 232
Charles II, King of England, 96
Chemical Warfare Service, U.S., 196
Chernin, Peter, 223
Chewa people, 123–27
Childs, St. Julien Ravenel, 56
Chile, 251*n63*
China, 48, 111–14, 117–19, 146–47, 239; ancient, 37, 223; Cultural Revolution in, 111–12; Military Academy of Medical Sciences, 113; Project 523 in, 112–13, 120; in World War II, 100
chloroquine, 102–104, 114, 126, 165, 222; parasites resistant to, 108–12, 114–16, 118, 120, 212, 217, 218, 220, 223, 239–40
cholera, 20, 24, 79, 143, 150, 192, 204
Christianity, 67, 137–38
Christophers, Samuel Rickard, 176
Church of Scotland, 56
cinchona bark, 89–96, 99, 102, 251*n63*, 260*n45*
Civil War, 73–74, 76
Clarkson, M. L., 197
Clinton, Bill, 223
Clinton Foundation, 228
Coartem, 115

Coatney, Robert G., 109
Cobbold, Thomas Spencer, 148–49
Coburn, Tom, 233
coffee, 91, *260n24*
cold war, 204, 214–15
Colin, Léon, 146
Colombia, 83, 108, 150, 217
Columbia University, 217
Columella, Lucius, 147
Comoros Islands, 120
Congress, U.S., 189, 207
Connecticut, 71–73, 75–76, 90
Conniff, Michael, 184
Consumer Reports, 215
Continental Congress, 90
Crichton, Michael, 233
Cromwell, Oliver, 90
crop rotation, 191, 200
Cuba, 150, 179
Culex mosquitoes, 155
Curtin, Philip, 38
Curtis, Chris, 171
Curtis, Richard, 225
Cushite people, 26

Dafra Pharmaceuticals, 117
Dante Alighieri, 68
DDT, 165, 193–212, 214, 222, 237; development of, 193–95; in eradication campaigns, 198–99, 202–209, 214, 219; free-market conservative advocacy of, 233; mosquitoes resistant to, 201, 205–206, 211–12, 217, 220, 227, *284n76*; toxicity and environmental impact of, 200, 209, 214, 215
Defoe, Daniel, 173
dengue fever, 150
dichlorodiphenyltrichloroethane, *see* DDT
diptheria, 192
DNA, 16, 21–22, 66, 133
Drake, Francis, 45
Drummond, Henry, 13, 14
Duffy antigens, 22, 24, 47
Durant-Reynals, M. L., 94, 100
dysentery, 39, 40, 52, 160

Earl, Ralph, 71–72
Earle, Carville, *249n19*
East African, The, 238
East India Company, 48, 49, 175
Ecuador, 83
Egypt, 197, 201, 227; ancient, 37
Eisenhower, Dwight D., 207
El Niño, 83–84
El Salvador, 201, 205
encephalitis, Japanese, 240
England, *see* Britain
Equatorial Guinea, 231, 233
Essig, E. O., 197
Etang, Josiane, 227
Ethiopia, 81, 115
ExxonMobil, 141, 231, 232

Facebook, 223
Fairley, Neil Hamilton, 104–109
falciparum malaria, 42, 52, 48, 87, 96, 106; artemisinin combination medications for, 237, *267n195*; in childhood, 31–33, 35, 133; chloroquine-resistant, 106–108, 110–11, 217, 218, 240; in Roman Empire, 65–68, 70; sickle-cell gene and, 28, 47; prophylaxis of, 87, 106; slave trade and spread of, 43, 45–48,

177; tolerance to, 36, 37; vaccine research for, 166–68
Falleroni, Domenico, 163
Family Health International, 122
Fansidar, 109, 114
Far Eastern Economic Review, 113
Farley, John, 213
farming, *see* agriculture
Fascists, 159, 197, 198
Febris, demon goddess of malaria, 64, 67, 69
filariasis, 146–49, 151
Fortune magazine, 94
France, 91, 95, 146, 151, 163, 180, 226; colonies of, 42, 144, 212; settlers in American colonies from, 47, 91; in World War I, 77
Fungladda, Wijitr, 136

Galen, 64
Gambia, 38, 124, 223, 226
Gandhi, Mohandas, 209, 210
Garibaldi, Giuseppe, 144
Garnham, P.C.C., 176
Gates, Bill, 223, 236
Gates, Melinda, 236
Gauguin, Paul, 150
Ge Hong, 112
Geigy Corporation, 194
Geisel, Theodor Seuss, 100–101
Genzyme, 141
Germany, 143, 146, 157, 158, 203; Mad Cow Disease in, 121; quinine in, 93, 98; in World War I, 78–79; in World War II, 99, 101, 102, 194, 196
Ghana, 89, 212, 229; University of, 114
Gladwell, Malcolm, 185
GlaxoSmithKline, 166, 171
Global Fund to Fight AIDS, Tuberculosis and Malaria, 115, 116, 139, 223, 224, 236, 238
Godfrey of Viterbo, 65
Goebbels, Joseph, 196
Gorgas, William Crawford, 178–85, 187, 188
Gorgas Memorial Institute, 5, 6, 130, 142–43
Grassi, Giovanni Battista, 154–55, 157–58, 160, 175
Greece, 77–79, 198, 199, 201, 205, 208; ancient, 37, 225
Greenwood, Brian, 234
G6PD, 63
Guadalcanal, 99, 194
Guangzhou University, 113
Guardian, The, 224
Guatemala, 124
Guyana, 212

Hackett, Lewis, 56, 58, 78, 163–65, 168, 197, 201, 225
Hand, Landrift, 187–88
Harper's magazine, 99, 137
Harper's Weekly, 150
Harrison, Gordon, 56, 57
Harrison, Mark, 175
Harugoli, Kishor, 82
Harvard University, 14, 166, 188, 189, 208, 216; Malaria Initiative (HMI), 141–43, 167–69
Hasted, Edward, 172–73
Havana, 179
Hayman, Martin, 135
Haynes, Douglas M., 148
Hecate, 66
Hedge Funds vs. Malaria, 223

Heggenhougen, H. Kristian, 136
Helitzer, Deborah, 123–24, 126, 127
hemoglobin E, 23–24
Herodotus, 225
Heston, Charlton, 196
Hippocrates, 37
Hiroshima, atomic bombing of, 196
Hitler, Adolf, 233
HIV, 129, 222
Ho Chi Minh Trail, 111
Hoffman-La Roche, 109
Holland, *see* Netherlands
Homer, 37
Hong Kong, 111
House of Representatives, U.S., 196
Howard, L. O., 162
Hudson Institute, 234
Human Genome Project, 142
Humphrey, Hubert, 207
Humphreys, Margaret, *249n19*
hydropower industry, 185–90

Ice Age, 21
I.G. Farben, 101
Illinois, 177
Illinois Central Railroad, 179
Illovo Sugar, 123
Imperial Chinese Maritime Customs Service, 146
Independent, The, 82
India, 3–4, 125, 164, 209, 216, 217; ancient, 37, 147; bed nets in, 226; British Raj in, 95, 149, 152–54, 157, 159, 165, 175–77; economic growth and building boom in, 81–82; El Niño and, 84; population density in, 59; during World War I, 77; during World War II, 99
Indian Medical Gazette, 158
Indian Medical Service (IMS), 152, 157, 159–60, 176
Indonesia, 83, 92, 206, 208, 212, *261n45*; in World War II, 99
influenza, 44; avian, 9; pandemic of 1918, 79
Innocent VIII, Pope, 68
insecticides, 139, 181, 195–97, 201, 218, 220, 232; bed nets treated with, 223–28; mosquitoes resistant to, 211, 217, 227, 235; *see also* DDT
intermittent preventive therapy for infants (IPTI), 234–35
International Development Advisory Board (IDAB), 203–208, 212, *284n76*
International Monetary Fund (IMF), 221
Interpol, 119
Iran, 205, 212
Ireland, settlers from, 46
Irian Jaya, 212
Islam, radical, 233
Italy, 15, 82, 143–44, 159, 163, 199, 208; Fascist, 160; quinine distribution in, 97–98, 239; research in, 151, 154–55, 157, 165; in World War I, 77; in World War II, 194; *see also* Rome

Jackson, Harvey H., 187
Jain religion, 4, 10
Jamaica, 53, 150
James, Sydney Price, 160
Jamestown, colonial, 39–40, *249n19*

Japan, 48, *260n45*; population density in, 59; in World War II, 99–100, 196
Japanese encephalitis, 240
Java, 92–94, 102, 164, *260n45*
Jefferson, Thomas, 57
Jesuits, 89–90
Jewish refugees, 104
Johns Hopkins University, 166, 178, 208
Justice Department, U.S., 94

Kennedy, John F., 207, 215
Kenya, 83–84, 217, 222, 227
Khoisan people, 36
Klebs, Edwin, 144, 145
Koch, Robert, 98, 143, 146, 157, 159
Kochi, Arata, 118, 235, 236
Kuna people, 6–8, 49, 52, 53, 218, 240
Kutcher, Ashton, 224

Laifer, Lance, 219, 223, 233
Lancet, The, 118, 149
La Niña, 83
Laveran, Alphonse, 144–46, 151
League of Nations, 160-62, 176
Lebanon, 201, 205
Ledger, Charles, 92
Ledgeriana trees, 99
Leinengen versus the Ants (radio play), 196
Lepes, Tibor, 216
LePrince, Joseph, 179–81, 188
leprosy, 44, 87
Lesseps, Ferdinand de, 149–51
Lewis, Timothy, 149
Liberia, 220
Li Guoqiao, 113, 120
Lines, Jo, 171
Litsios, Socrates, 134, 228–29
Little Ice Age, 83
Liverpool School of Tropical Medicine, 185
Livingstone, David, 96, 97, 136–39
Livingstone, Mary, 96
London Daily Mail, 185
London School of Hygiene and Tropical Medicine, 165, 170–71, 202
London Society of Arts, 159
Long, Richard, 53
Louisiana, 57; during Civil War, 73–74
Lydus, John, 68–69
Lyle, Clay, 197

MacArthur, General Douglas, 99, 100, 104
Macaulay, Lord Thomas Babington, 50
Macedonia, 76–79
Madagascar, 35, 58
Mad Cow Disease, 121
Malaria No More, 223–25, 236
Malawi, 11–12, 29–33, 86–87, 131, 132, 137–39, 239, *270n44*; economic impact of malaria in, 230; traditional village culture in, 122–28
Malaysia, 131, 160, 164, 209, 239
Mamani, Manuel Incra, 91–93
Mande languages, 26
Man's Mastery of Malaria (Russell), 202
Manson, Patrick, 146–49, 151–59, 162, 165
Mao Zedong, 111
Marathon Oil, 56, 231, 232

Marshall Plan, 207
Martini, Erich, 194
Martinique, 150
Maryland, colonial, 41
Massachusetts, 73, 76
Massachusetts General Hospital, 131
McCutchan, Tom, 239
McElwee, Pamela, 111
measles, 12, 20, 24, 34, 39, 42, 44, 134
Médecins Sans Frontières, 115–16
Medical Times and Gazette, The, 137
Medicines for Malaria Venture, 119
Mendis, Kamini, 13
Metropolitan Museum (New York), 72
miasmatic theory, 43–44, 143, 146, 179
Michigan, 89, 177
Michigan, University of, 213
Michigan State University, 214
Middle Ages, 37, 44
milldams, 75–76
Mills, Anne, 221
Mindanao, 99
mining, 79–81, 231–32
Missionary Travels and Researches in South Africa (Livingstone), 137
Mohéli Island, 120
Morocco, 208
Mosquirix, 166
Mosquitoes, Malaria, and Man (Harrison), 56
Motta, Jorge, 130
Mozambique, 96, 135, 227, 231–32
Msechu, June, 125
Multilateral Initiative on Malaria, 34
mululuza, 88, 89
Muppets, the, 225
Mussolini, Benito, 160
Myanmar, 115, 118, 229

Nájera, José, 102, 211
Namibia, 226
Naples, 68
National Academies, Institute for Medicine of, 116, 125
National Cancer Institute, 142
National Malaria Society, 201
Native Americans, 39, 40, 42
Nature (journal), 168
Navy, U.S., 111
Ndoye, Tidiane, 128
Netherlands, 158, 160, 191, 231; Kina Bureau, 94, 99, 100, 102–103, 260*n45*; colonies of, 42, 92–95
New England, 74; colonial, 41, 50, 70
New Guinea, 84, 98, 99, 105–108, 194, 207
New Jersey, 58, 71, 74, 161
Newman, Peter, 213
New Mexico, University of, 123
News Corporation, 223
Newsweek, 215
New Yorker, The, 185
New York Times, The, 74, 75, 83, 99, 100, 135, 136, 146, 184, 194, 195, 200, 205, 207, 224, 225, 232, 236
Nicaragua, 120
Nigeria, 117, 206, 220; Ministry of Health, 238
Nightingale, Florence, 69, 176
Nobel Prize, 158
non-governmental organizations (NGOs), 223
Nothing but Nets, 223
Nott, Josiah Clark, 57
Novartis, 114–16, 139–40, 171

Nubian people, 37
nuclear weapons, 214

oil companies, 56, 141, 171, 231, 232
On the Beach (film), 215
Osler, William, 185
Owen, H. Collinson, 78
Oxford University, 133–34

Packard, Randall, 212
Palestine, 77
Panama, 5–10, 119, 130, 142–43, 199, 206, 240; canal building projects in, 5, 149–51, 178–85, 187; Health Department of, 218; Scottish settlers in, 48–55, 93; Spanish colonization of, 45–46, 48, 49, 53, 55
Pan American Health Organization (PAHO), 199, 203, 216
Pan American Sanitary Bureau, 199
Pan American Sanitary Conference (1954), 202
Papua New Guinea, 84, 99, 207
parasitism, 13–14
Paris Green, 197
Paterson, William, 48–51, 53, 56
Peace Corps, 124
Pécoul, Bernard, 115–16
Pelletier, Pierre, 91, 95, 102–103
People in Red, 236
Perkins, John, 200
Peru, 81, 83, 94
pesticides, *see* DDT; insecticides
petroleum industry, 56, 141, 171, 231, 232
Philippines, 98–100, 124, 212
Pilgrims, 40
plague, 9, 34, 44, 143
Plasmodium, 5, 17–30, 88–89, 121, 132, 144, 180, 184, 192, 216, 241; in ancient world, 37; artemisinin and, 116, 117; DDT and, 193, 202, 220; evolution of, 13–14, 239; genetic diversity of, 103; identification of mosquito as carrier of, 154; microscopic discovery of, 145, 151; morphology and physiology of, 15; mosquito behavior and, 59; plants alkaloids and, 88–89 (*see also* cinchona bark); quinine and, 89, 95, 99, 220; research on vaccine against, 167; slave trade and, 41
species: *P. falciparum*, 7, 8, 27–29, 32, 33, 35–38, 40, 57, 62, 65–66, 68, 70, 73, 77, 81, 110–11, 132–33 (*see also* falciparum malaria); *P. knowlesi*, 239–40; *P. malariae*, 20–21, 23, 25, 27, 77, 167; *P. ovale*, 25, 68, 132, 167; *P. vivax*, 7, 8, 21–24, 26–27, 37, 40, 47, 61–62, 68, 70, 77, 102, 132, 167, 172, 177
Plymouth colony, 43
pneumonia, 114, 183
pneumonic plague, 44
Politico.com, 228
Poliziano, Angelo, 69
polymerase chain reaction (PCR), 133
Popular Mechanics, 196
populism, American, 184
Portugal: colonies of, 42, 96; missionaries from, 38
Powell, Nathaniel, 39, 40
Prebble, John, 50
Protestants, 90
Puritans, 40
Pygmy people, 36
pyrethrum, 181, 182, 227

Queen Elizabeth Hospital (Blantyre, Malawi), 29–30, 128, 131, 132
quinacrine, 101–102, 104–108
quinine, 74, 86, 89–103, 111, 119, 137, 168, 182; artemisinin compared with, 112; for cerebral malaria, 129; effect on *Plasmodium* of, 89, 145; obstacles to widespread use of, 93–98; prophylactic use of, 86, 96–98, 239; side effects of, 97, 188; synthetic, *see* chloroquine; quinacrine; during World War II, 98–101

Raleigh, Walter, 173
Reed, Major Walter, 179
Republican Party, 233
Revolutionary War, *see* American Revolution
rheumatic fever, 87
Rhodesia, 80
Riamet, 114, 115
Ritchie, Thomas, 238
Rocco, Fiammetta, 93
Rochester City Newspaper, 232
Rockefeller, John D., 161
Rockefeller Foundation, 161, 163, 165–66, 191, 197–200
Rohwer, Sievert, 195
Roll Back Malaria (RBM), 223–24, 227, 229, 236, 238–39
Roman Catholic Church, 67, 89–90
Rome, 68–70; ancient, 61–68, 70, 147, 172
Roosevelt, Franklin D., 190
Roosevelt, Theodore, 179, 182, 184
Rosenberg, Tina, 136
Ross, Ronald, 152–60, 162, 175, 225
Roubaud, Emile, 163
Royal Air Force, 209
Royal Dutch Shell, 171
Royal Society, 149
Royal Society of Tropical Medicine and Hygiene, 211
Ruggles, Joseph, 72, 76
Ruskin, John, 69
Russell, Paul, 56, 98–100, 198–200, 202–208, 212, 214, 216, 217
Russia, 241; in World War I, 79

Sabot, Oliver, 228
Sacculina carcini, 13
Sachs, Jeffrey, 223, 230, 236
Sammonicus, Serenus, 64
Sanaria, 166
Sanofi-Aventis, 116–17, 238
Santo Spirito hospital (Rome), 68
Sarawak, 212
Sardinia, 65, 68, 164, 198, 199, 206, 208, 213, 231
Sarrail, General Maurice, 78
SARS, 9
Saudi Arabia, 201, 205–206
Science magazine, 168, 194
Scotland, 48–49, 56; medical education in, 148; settlers in colonial America from, 46, 49–56, 93
Senegal, 114, 133; Research Institute, 128
September 11, 2001, terrorist attacks, 232
Seuss, Dr., 100–101
Seydel, Karl, 270*n44*
Shortt, H. E., 176
Sicily, 65, 68, 144, 194
Silent Spring (Carson), 215
Simmons, James Stevens, 45, 183, 200

Singapore, 82
Sixtus V, Pope, 68
slaves: African, *see* African slaves; in ancient Rome, 65; Native American, 41–42
sleeping sickness, 38
smallpox, 9, 12, 20, 24, 34, 39, 42, 44, 204; eradication campaign against, 216, 217; vaccines against, 166–67
Smith, John B., 161
Snowden, Frank, 98, 70
Solomon Islands, 99
Soper, Fred, 197–99, 206, 216
Soren, David, 65–66
South, University of the, 178, 185
South Africa, 227
Soviet Union, 204, 207; nuclear weapons testing in, 215
Spain, 164; American colonies of, 39, 42, 45–46, 93
Spearpoint, C. F., 80
Spice Islands, 48
Spielman, Andrew, 188, 189, 216, 217, 229, 230
Sri Lanka, 84, 198, 199, 208, 213, 215–16, 231, 238
State Department, U.S., 204, 207 (*see also* International Development Advisory Board)
Steketee, Rick, 115
Stendhal, 70
Sternberg, George Miller, 145–46
Stevens, General J.E.S., 105–106
Straits Times, 82
Sudan, 227, 232
Suez Canal, 149, 150
sulphur insecticides, 181, 182
Sumatra, 164
Sumeria, 37
Suriname, 212
Swellengrebel, N. H., 160
Switzerland, 194; Tropical Institute of, 125
Sydenham, Thomas, 172

Taft, William Howard, 182
Taiwan, 203
Tajikistan, 8
Talbor's Wonderful Secret, 90
Tanzania, 125, 128, 212, 217
Taylor, Norman, 93
Taylor, Terrie, 11–13, 29, 30, 32–33, 86–87, 129, 132, 139
Tennessee Valley Authority (TVA), 190–91
terrorism, 232
tetanus, 143
Texas, 179
Thailand, 108–109, 115, 118, 136, 217, 239, 267*n195*
Them! (film), 196
Time magazine, 196
Tommasi-Crudeli, Corrado, 144–46
Torres Strait Islands, 84
Townshend, Charles, 191
tuberculosis, 44, 79, 134, 143, 160, 192
Turkey, 8, 216
Twitter, 224
typhoid, 39, 40, 44, 52, 55, 143, 192, 249*n19*
typhus, 44, 52, 55, 192

Uganda, 220
United Nations, 223, 234, 236; Children's Fund (UNICEF), 115–16, 171, 203, 215; Food and

United Nations (*cont.*)
Agriculture Organization, 214; Relief and Rehabilitation Administration (UNRRA), 198
United States, 145, 166, 177–91, 221, 228, 238, 240–41, 260*n24*; antitrust suit against Kina Bureau by, 194; DDT in, 195–201, 211, 213–15, 220; endangered species in, 279*n86*; entomological research in, 160–62; eradication campaign funded by, 203–208, 212, 215, 217; history of malaria in, 56–57, 70–76, 177–78, 185–92; nuclear weapons testing in, 215; and Panama Canal, 151, 178–85, 187; public health service of, 186, 201; racism in, 57–58; support for international antimalarial projects from, 223–25, 227–28, 232–36, 238; in Vietnam War, 111; in World War II, 98–102, 191, 194–96; *see also specific government agencies and departments; specific states*
U2, 223

vaccines, 15, 166–67
Vargas, Gertülio, 197
Varro, Marcus Terentius, 63
Vatican, 68–69
Vedic sagas, 37
Venezuela, 84, 150, 198, 217
Veto the 'Squito, 223
Viele, General Egbert L., 75, 76
Vietnam, 117; war in, 110–13, 209, 210
Vileisis, Ann, 190
Virginia, 178 colonial, 39–41
Visigoths, 67
Vogt, William, 214

Wafer, Lionel, 49–52, 54, 251*n63*
Wall Street Journal, The, 219, 233
Walpole, Horace, 69
Washington, George, 90
Washington Monument, 179
Washington Post, The, 233
Watson, Malcolm, 80
Webb, James, 215
Welch, S. W., 188–89
West Nile virus, 240
West Point, U.S. Military Academy at, 178
Whorton, James, 195
Wirth, Dyann, 142, 167–68
Wisconsin, 177
World Bank, 167, 221, 223
World Health Assembly, 203, 216
World Health Organization (WHO), 102, 125, 138, 199, 201, 209, 218, 228, 286*n103*; and AIDS crisis, 222; artemisinin policies of, 113, 115, 117, 118, 238; on cerebral malaria, 129; and chloroquine resistance, 109–10; environmental impacts of economic development cited by, 79; eradication campaigns of, 202–205, 211–12, 215–17, 219–20; establishment of, 108; estimates of worldwide malaria deaths by, 134; on intermittent preventive therapy for infants, 234–35; Roll Back Malaria and, 223, 229, 236; and vaccine research, 167
World Malaria Day, 232
World Swim Against Malaria, 223
World War I, 76–79, 93, 97–98, 175

World War II, 98–102, 104–108, 115, 191, 193–96

yellow fever, 46, 51–52, 55, 79, 150, 151, 179–80, 182, 183; vaccine against, 166–67

Zaire, 221
Zambia, 221, 233
Zanzibar, 226
Zgambo, Joseph, 80
Zhou Yiqing, 112, 113
Zimbabwe, 227
Zimmer, Carl, 13